FUELS AND ENERGY FROM RENEWABLE RESOURCES

Edited by

David A. Tillman
Materials Associates, Inc.
Washington, D.C.

Kyosti V. Sarkanen
College of Forest Resources
University of Washington
Seattle, Washington

Larry L. Anderson
Department of Mining, Metallurgical and Fuels Engineering
University of Utah
Salt Lake City, Utah

ACADEMIC PRESS
New York San Francisco London 1977
A Subsidiary of Harcourt Brace Jovanovich, Publishers

ACADEMIC PRESS, INC.
111 Fifth Avenue, New York, New York 10003

United Kingdom Edition published by
ACADEMIC PRESS, INC. (LONDON) LTD.
24/28 Oval Road, London NW1

Library of Congress Cataloging in Publication Data

Symposium on Fuels and Energy from Renewable Resources, Chicago, 1977.
Fuels and energy from renewable resources.

Includes index.
1. Renewable energy sources–Congresses.
2. Fuel–Congresses. I. Tillman, David A.
II. Sarkanen, K. V., Date III. Anderson, Lary LaVon. IV. American Chemical Society.
V. Title

TJ163.2.S96 1977 621.4'023 77-13450
ISBN 0-12-691250-5

PRINTED IN THE UNITED STATES OF AMERICA

FUELS AND ENERGY FROM RENEWABLE RESOURCES

Academic Press Rapid Manuscript Reproduction

SYMPOSIUM ON FUELS AND ENERGY FROM RENEWABLE RESOURCES

174th National Meeting of the American Chemical Society, Chicago, August, 1977

CONTENTS

LIST OF CONTRIBUTORS

Numbers in parentheses indicate the pages on which authors' contributions begin.

J. A. Alich, Jr. (213), Stanford Research Institute, 333 Ravenswood Avenue, Menlo Park, California 94025

Larry L. Anderson, Department of Mining, Metallurgical and Fuels Engineering, 309 Mineral Science Building, University of Utah, Salt Lake City, Utah 84112

Richard P. Arber (275), CH2M Hill, Inc., 12000 E. 47th Avenue, Denver, Colorado 80239

William Arlington (249), Florida Sugar Cane League, Inc., P.O. Box 1148, Clewiston, Florida 33440

David L. Brink (141), College of Natural Resources, University of California, Berkeley, California 94720

John Burford (275), Bio-Gas of Colorado, 342 E. Third Street, Loveland, Colorado 80537

William F. DeGroot (93), Wood Chemistry Laboratory, University of Montana, Missoula, Montana 59812

L. W. Elston (169), Engineering Experiment Station, Georgia Institute of Technology, Atlanta, Georgia 30332

R. K. Ernest (213), Stanford Research Institute, 333 Ravenswood Avenue, Menlo Park, California 94025

George W. Faltico (141), Kaiser Engineers, Oakland, California 94612

John B. Grantham (55), Pacific Northwest Forest and Range Experiment Station, 1601 Second Avenue Building, Seattle, Washington 98101

James R. Greco (289), National Solid Wastes Management Association, 1120 Connecticut Avenue, NW, Suite 930, Washington, D.C. 20036

Richard H. Hamilton (213), Stanford Research Institute, 333 Ravenswood Avenue, Menlo Park, California 94025

D. R. Hurst (169), Engineering Experiment Station, Georgia Institute of Technology, Atlanta, Georgia 30332

L. N. Johanson (197), Department of Chemical Engineering, University of Washington, Seattle, Washington 98195

J. A. Knight (169), Engineering Experiment Station, Georgia Institute of Technology, Atlanta, Georgia 30332

William F. Lalor (257), Cotton Inc., 4505 Creedmoor Road, Raleigh, North Carolina 27612

B. M. Louks (213), Stanford Research Institute, 333 Ravenswood Avenue, Menlo Park, California 94025

Thomas R. Miles (225), P.O. Box 216, Beaverton, Oregon 97005

K. A. Miller (213), Stanford Research Institute, 333 Ravenswood Avenue, Menlo Park, California 94025

Charles M. Mottley (1), Fossil Energy Division, U.S. Energy Research and Development Administration, 20 Massachusetts Avenue, NW, Washington, D.C. 20545

Kyosti V. Sarkanen (169), College of Forest Resources, University of Washington, Seattle, Washington 98195

M. D. Schlesinger (313), 4766 Wallingford Street, Pittsburgh, Pennsylvania 15213

F. A. Schooley (213), Stanford Research Institute, 333 Ravenswood Avenue, Menlo Park, California 94025

Fred Shafizadeh (93), Wood Chemistry Laboratory, University of Montana, Missoula, Montana 59812

Jerome F. Thomas (141), College of Engineering, University of California, Berkeley, California 94720

David A. Tillman (23), Materials Associates, Inc., 600 New Hampshire Avenue, NW, Suite 960, Washington, D.C. 20037

Frederick T. Varani (275), Bio-Gas of Colorado, 5620 Kendall Court, Unit G, Arvada, Colorado 80002

T. C. Veblen (213), Stanford Research Institute, 333 Ravenswood Avenue, Menlo Park, California 94025

George D. Voss (125), American Fyr-Feeder Engineers, 1265 Rand Road, Des Plaines, Illinois 60016

J. G. Witwer (213), Stanford Research Institute, 333 Ravenswood Avenue, Menlo Park, California 94025

John I. Zerbe (115), U.S. Forest Products Laboratory, P.O. Box 5130, Madison, Wisconsin 53705

PREFACE

Fuels and Energy from Renewable Resources, as both a symposium and a volume, stemmed from the 1976 American Chemical Society Symposium, *Thermal Uses and Properties of Carbohydrates and Lignins* (also published by Academic Press). The discussions held at San Francisco concluded that serious energy planning should begin, that it should not overlook biomass resources such as silvicultural and agricultural residues, and that these resources should be considered for the near and mid term as well as the long term. Renewable resources, particularly residues, can aid the United States in getting to the year 1985 and beyond.

The symposium Fuels and Energy from Renewable Resources was held at the 1977 Fall Meeting of the American Chemical Society. It was sponsored by both the Cellulose and Fuel Divisions. It addressed the issues of energy planning, and the incorporation of crop and conifer residues into the development of energy supply.

To set the framework for the discussions, two papers were given at the beginning—one forecasting total U.S. energy needs and the other depicting the present energy contributions of nonfossil organic materials. From there the symposium moved into a detailed discussion of silvicultural materials—the volume potentially available, their fuel value, and methods of utilization. The energy production–conservation system of generating both steam and electricity from pulp and paper mill residues was among the near-term concepts gaining attention. Limitations on wood as an energy source were also discussed. Similarly, the symposium focused on agricultural residues including contributions on collection, combustion, and conversion. Urban waste concluded the areas of concern covered at this meeting.

The symposium, as a whole, was a multidisciplinary effort. In addition to the chemical and chemical engineering questions of fuel value, conversion, and utilization, issues of economics, environmental protection, and institutional impediments were also addressed. The papers presented at the symposium and in this volume reflect a broad diversity of skills brought to bear on this single aspect of the energy situation.

It is useful to note that, while we are rediscovering renewable resources, other countries regularly include them as part of their energy supply system. Sweden, for example, gets 8% of its energy from wood, while Finland gets 15% of its energy from that source and Brazil gets 27% of its energy from forest fuels. The People's Republic of China has 100,000 operational gas producers, which convert manure and crop residues into methane-rich gas by anerobic digestion. India has 50,000 such digesters and South Korea has 20,000. Now this country, driven by limited supplies of oil and gas, is moving in this renewable resource direction.

It is also important to observe that renewable resources, while they can play a far more significant role than they do, will always be a supplemental fuel source. Their potential, at least for the near term, appears to be in the eight quadrillion Btu (quad) per year range. The economy as a whole consumes about 75 quads annually, and that rate of consumption is expected to rise in the coming years. Still, eight quads is a lot of energy. It is equivalent to 1.4 billion barrels of oil, which—when imported—currently cost $18 billion landed in the U.S.

In order to approach that potential, however, we must first define how much energy we really need; how much we are getting at the present time; and what technologies are now available, or will be available shortly, to improve upon the present contribution from renewable resources. To that end this symposium and this volume were established, and this segment of the energy discussion and debate was joined.

The editors would like to acknowledge the cooperation of all contributors in their timely preparation of excellent papers. We would also like to acknowledge the assistance of Mrs. Mildred Tillman, who in addition to assisting the co-chairmen in the preparation and mailing of letters and forms, typed all of the papers for publication.

David A. Tillman
Kyosti V. Sarkanen
Larry L. Anderson

HOW MUCH ENERGY DO WE REALLY NEED*

Charles M. Mottley

Office of the Assistant Administrator for Fossil Energy
U.S. Energy Research and Development Administration
Washington, D.C.

I. INTRODUCTION

The question of how much energy does the Nation need has been debated for several years. The President's National Energy Plan issued last April has focused attention on the energy supply and demand situation for the next eight years. However, we need quantitative estimates of requirements for the longer term to guide the energy research and development effort. A rationale called the requirements approach is developed in this paper and offered as a way to establish strategic objectives for that purpose.

---

* The conclusions and opinions expressed in this paper are those of the author and are not to be considered as statements or positions of the U.S. Energy Research and Development Administration.

The requirements approach has not been used in energy planning and budgeting.* The procedure used most frequently is to identify existing trends in energy supply or demand and then attempt to trace a likely future using a chain-of-events or scenario technique. The forces of the market place and prices figure prominently in the scenario approach. However, one of the difficulties is that in the long run the outcomes are strongly influenced by the unpredictability of prices. Furthermore, because a very large number of variables are involved, a host of alternate futures can be generated. In practice there is usually no logical ground offered to the decision maker for choosing among scenarios; thus bias is apt to condition the selection.

Because the scenario approach has an open end and generates a large number of possible outcomes, it fosters disagreement and uncertainty about strategic objectives. The requirements approach on the other hand has a closed end. It starts by setting the objectives. This converges agreement and aids in the selection of the preferred ways and means to do the job. When objectives have been set, then scenarios can be used properly to test the desirability and feasibility of any proposed course of action designed to achieve those objectives. Circumstances change in the real world. Therefore, ways must be left open, with enough lead time, to change the targets. For that reason policy formulation based on the requirements approach must be coupled with contingency planning.

It is evident from estimates of the situation that all the potential sources of energy, including renewable resources, must be considered as we determine our future requirements. Given credible estimates of total requirements, it is possible using

---

* The concept is not mentioned as a viable alternative in two recent papers, issued by the Congressional Budget Office: (1) Energy Policy Alternatives and (2) Energy Research, Development Demonstration and Commercialization, Washington, D.C., January 1977.

appropriate analytical techniques to disaggregate the total into individual targets for specific fuels, consuming sectors and regions. This paper is limited to estimation of total requirements. The rationale for estimating the total or aggregate national energy requirements is presented in the next section. In the subsequent section the rationale is used to frame quantitative energy target areas. The effect of conservation on the proposed targets is also examined.

II. THE RATIONALE FOR ESTIMATING ENERGY REQUIREMENTS

The rationale for estimating future energy requirements is based on the obvious proposition that people need and use energy. Fairly good estimates are available regarding the size of the U.S. population for the next 25 to 30 years. For example, the expected number of 16-year-olds (i.e., those young people becoming available for entry into the labor force) can be estimated within known limits to the year 1993. They are already born. Actuarial calculations tell how many will survive. Incidentally, the number of people reaching age 16 will be decreasing at the rate of about 70,000 per year for the next 15 years. This has important implications for the development of the rationale.

The population projections published recently by the Bureau of the Census [1] provide new estimates based on lower fertility rates. The projections are reported in three "Series." The first, Series I, is similar to the one on which several current energy forecasts have been based, but it is now regarded as being much too high. The other two, Series II and III shown in Table I, appear to be more realistic. Because of the lower fertility rates now occurring in the U.S., it seems reasonable to select Series II as an upper boundary and Series III as a lower boundary for estimating the size of the future U.S. population.

The table also shows the net population increase at 5-year intervals. Note that in both series the net increase declines after 1985.

TABLE I. Population Estimates and Projections (in millions of persons)

	Series II		Series III	
Year	Population	Net increase	Population	Net increase
1975	213.4	-	213.4	-
1980	222.8	9.4	220.4	7.0
1985	234.1	11.3	228.4	8.0
1990	245.1	11.0	235.6	7.2
1995	254.5	9.4	241.2	5.6
2000	262.5	8.0	245.1	3.9
2005	270.4	7.9	247.9	2.8

If the expected size of the future population and the amount of energy each person is expected to consume are known, then it might be a simple matter to compute the total energy requirements. A study of the situation reveals that it is not quite that simple, even though per capita consumption figures are often used for projection purposes.

Available data on gross energy consumed and total population for the 29 years, 1947-1975, are plotted in Fig. 1. The relationships between energy consumed and population was linear from 1947 through 1962, when an abrupt upward change occurred. A linear trend again prevailed to 1975. In the equations fitted to the data, population is expressed in millions and gross energy consumption in Quads (Btu × 10^{15}). The regression coefficients in the equations define the linear trends. In the first period

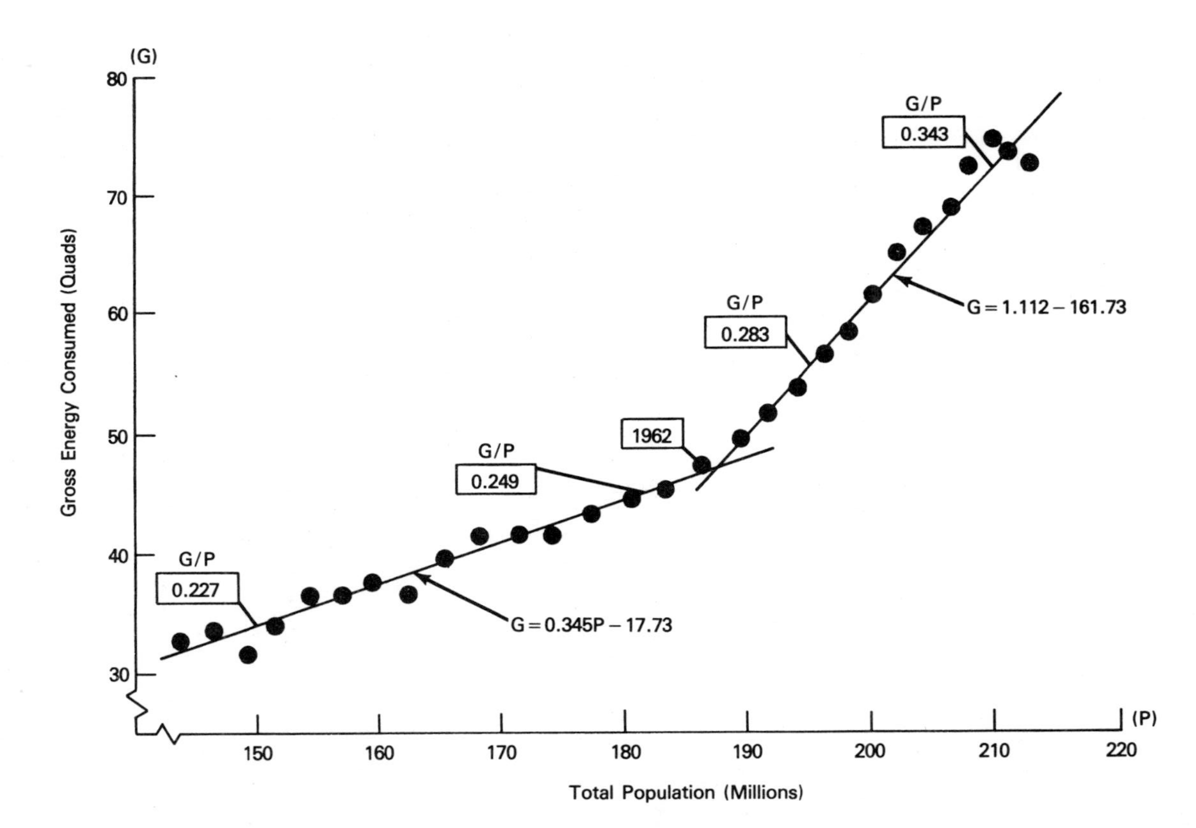

FIGURE 1. Energy consumed by the total population.

(1947 through 1962) each person added to the total population consumed 345 million Btu; after 1962 each additional person consumed 1112 million Btu. There appeared to be a sudden threefold increase beginning in 1962. However, if the ratios for the energy consumed per capita (G/P) are calculated, they would appear to grow as the series progresses: from about 227 million Btu per capita in 1950 to 343 million Btu in 1973. Note also that G/P ratios are much lower than the true rates as indicated by the linear regression coefficients.*

The question remains: Was there a sudden threefold increase in the rate of energy consumption beginning in 1962? The articulation of this question raises the possibility that the increase might be due to a change in the size of the work force. The growth of civilian employment is shown in Fig. 2. The slope of the two plotted lines represents the net increase in civilian employment. Each year new workers are added and others leave; the net result is an employed work force which has been growing for 30 years. During the period from 1947 through 1962, 637,000 people were added to the number of civilians employed each year. After 1962 the rate of growth was also linear but at the rate of 1,520,000 people per year, or about 2.4 times the original rate.

A more familiar descriptive statistic is the proportion of the total population comprised by the employed force. These data are shown in Fig. 3. Note the declining percentage from 1947 through 1962 and the rise from 1962 to 1974. Incidentally, the percentage of civilians employed in the U.S. has never been higher than 41%.

Because the ratios expressing the number employed as a percentage of the total population present the same mathematical

---

* The apparent growth of the G/P ratios is a mathematical anomaly; the line relating the two variables does not originate at zero and the negative intercept parameter of the equation has not been allowed for in the computation. Under such conditions ratios cannot validly be used for projections or comparative purposes.

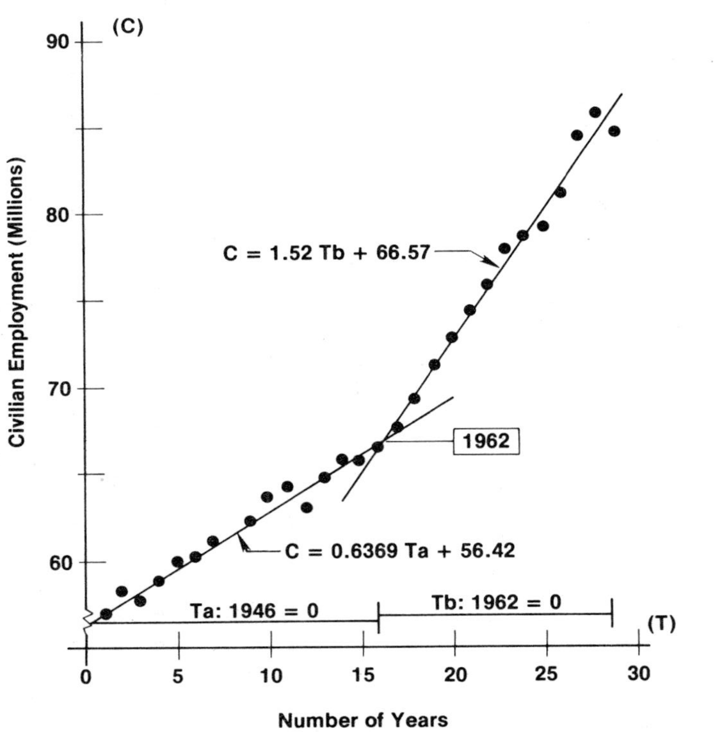

FIGURE 2. The growth of civilian employment.

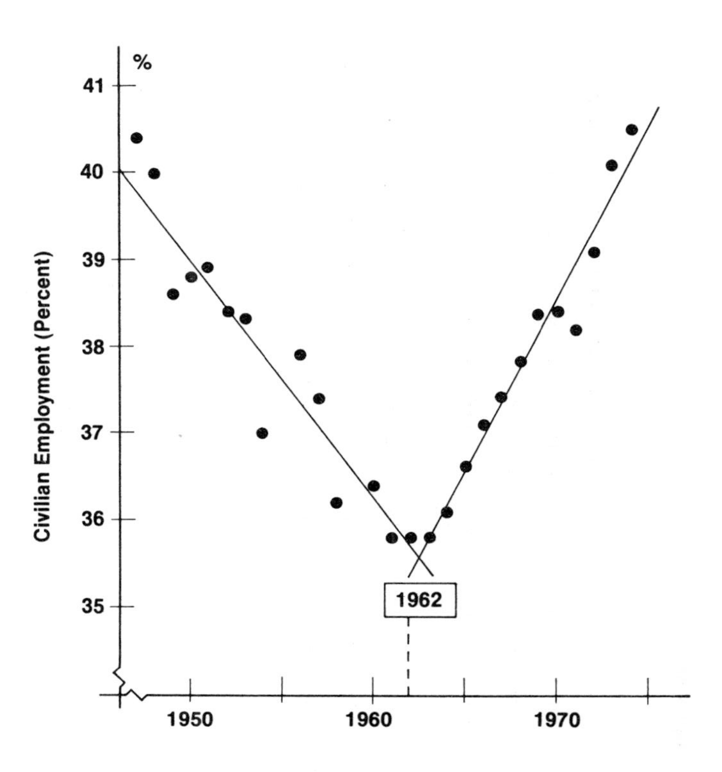

FIGURE 3. Civilian employment as a percentage of the total population.

difficulties as the G/P ratios discussed above, it is better to express the increase in the number of employees as a rate related to the increase in the total population. Accordingly, the linear regressions were fitted to the data as shown in Fig. 4. Note that in the first period (1947 to 1962) civilian employment grew at the rate of 224,000 per million increase in the total population; in the second period the civilian employment grew at the rate of 764,000 per million increase in the total population.

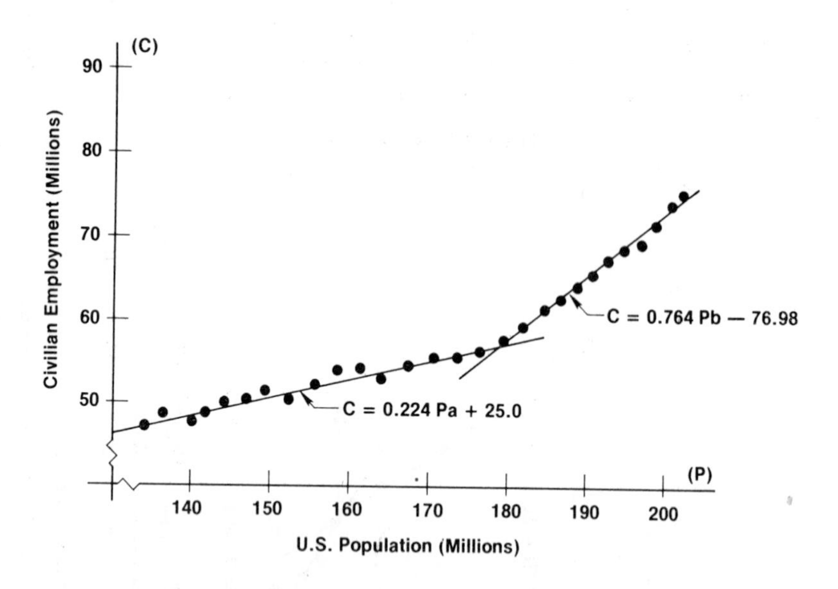

FIGURE 4. Civilian employment in relation to the total U.S. population.

The numerical relation of the employed civilian work force to the rest of the population is shown in Fig. 5. Note that for each additional employed person the rest of the population grew at the rate of 3.3 persons in the first period (until 1962) and at the rate of only 0.31 persons after that. This finding, as

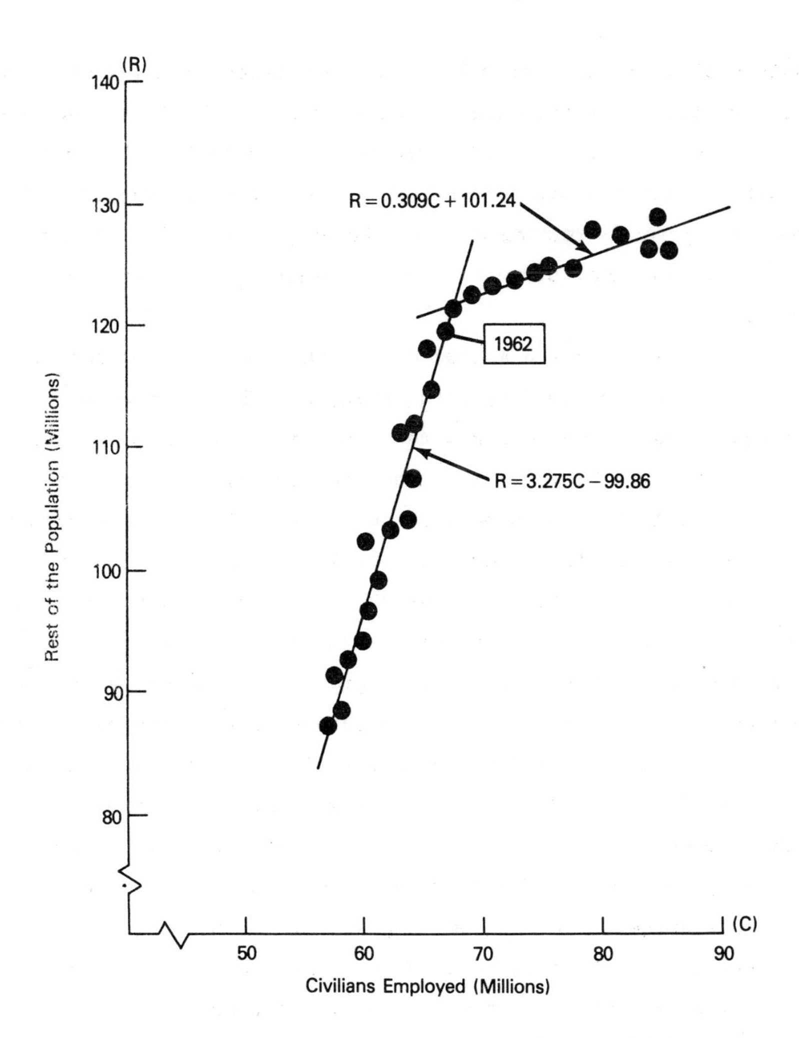

FIGURE 5. Relation of the rest of the population to civilian employment.

well as the data shown in Fig. 4, supports the conclusion that the increase in the size of the employed force reflects the transfer of young people resulting from the Post World War II baby boom from dependent status in the rest of the population to an employed status. The two rates of growth for the employed force shown in Fig. 4 represents two extremes. The first, at 224,000 additional employees per million of population from 1947 to 1962, reflects the input from the low birth rate in the Great Depression. The second, at 764,000 per million represents the input resulting from the high birth rate following World War II. These two rates provide reasonable, low and high boundaries respectively, for estimating the future size of the employed civilian work force.

It now becomes a matter of finding out how much energy is expected to be consumed by each employee. The relationship between gross energy consumed and the number of employed civilians is shown in Fig. 6. Note that the relationship is linear over the whole 29-year span with no break in 1962. Each person added to the work force accounts for 1.55 billion Btu of gross energy consumed. Note also that the value of the intercept parameter is negative and relatively large (-56.79). If G/C ratios were calculated they would show progressively increasing values over the 29-year span. This increase has been erroneously attributed to a change in individual lifestyle, when in fact an additional worker in 1974 consumed the same amount of energy as a worker added to the force in 1947. The additional total amount of energy consumed is due to more workers, not more energy consumed per worker. This is another way of saying that our increasing life style is related to higher employment.

The relationship between energy consumed, the employed force, and the rest of the population has been explored also. A multiple regression model was used to relate gross energy consumed (G), the number of civilians employed (C) and the rest of the population (R). The equation is as follows:

$$G = 1.50\ C + 0.066\ R - 60.88 \qquad (1)$$

Note that the partial regression coefficient for the employed force (C) is much greater than that for the rest of the population (R). Each additional nonworker would account for only 70 million Btu per year. The latter amount is so small that it

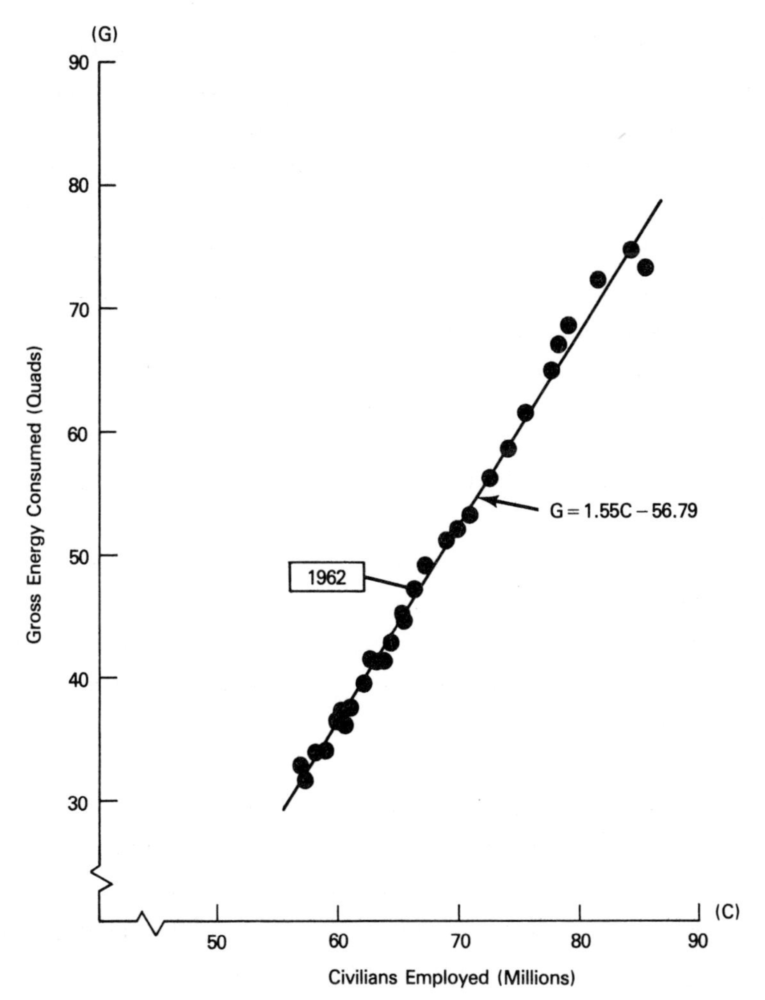

FIGURE 6. Relation of gross energy to civilian employment.

was decided to omit it and use the equation shown in Fig. 6 for estimation purposes.

$$G = 1.55\ C - 56.79 \qquad (2)$$

In summary, we now have the major quantitative factors for estimating future energy requirements: (1) Population projection boundaries within which the future size of the population can be expected to lie, i.e., between Series II and Series III; (2) estimates of high and low rates of addition to the employed civilian work force within which the number of employees can be expected to occur, i.e., between 224,000 and 764,000 per million of population increase; and (3) an estimate of the amount of energy consumed by each worker, i.e., 1.55 billion Btu. (If the current drive to achieve greater efficiency and conservation of energy use succeeds, then this consumption factor will have to be revised downward.)

III. ENERGY TARGETS OBTAINED

The basic population forecasts are shown in Table I.* The table also shows the net increase in population at 5-year intervals. To derive estimates of civilian employment the two equations shown in Fig. 4 were applied to the net population increase data. These data for the two series and the low and high rates of addition of employment at 5-year intervals out to the year 2005 were calculated. The results appear in Table II. The same data were used to compute the percentage of the total population represented by civilian employment shown in Table III. The gross energy requirements shown in Table IV, based on the employment estimates, were computed using equation (2), above.

By combining the data in Tables III and IV, the data for plotting the trapezoidal figures in Fig. 7 were derived. The trapezoids represent the area within which the targets for future energy requirements may be expected to lie. The centroids of the trapezoids were calculated and were taken as provisional targets.

---

* The data are derived from the Population Estimates and Projections published by the Bureau of the Census, February 1975. It should be consulted for a discussion of the methodology.

TABLE II. Civilian Employment (Millions)

Year	Series II Rate of Addition Low[a]	Series II Rate of Addition High[b]	Series III Rate of Addition Low[a]	Series III Rate of Addition High[b]
1975	85[c]		85[c]	
1980	87.1	92.2	86.6	90.3
1985	89.6	100.8	88.4	96.4
1990	92.1	109.2	90.0	101.9
1995	94.2	116.4	91.3	106.2
2000	96.0	124.4	92.2	109.2
2005	97.8	130.4	92.8	112.9

a. 224,000 employees/million net population increase.

b. 764,000 employees/million net population increase.

c. From observed data.

TABLE III. Civilian Employment (Percentage)

Year	Series II Rate of Addition Low	Series II Rate of Addition High	Series III Rate of Addition Low	Series III Rate of Addition High
1975	39.8[a]		39.8[a]	
1980	39.1	41.4	39.3	41.0
1985	38.3	43.1	38.7	42.2
1990	37.6	44.6	38.2	43.3
1995	37.0	45.7	37.8	44.0
2000	36.6	47.4	37.6	44.5
2005	36.2	48.2	37.4	45.5

a. From observed data.

TABLE IV. Energy Requirement (Quads)

	Series II		Series III	
	Rate of addition to civilian employment		Rate of addition to civilian employment	
Year	Low	High	Low	High
1975	75[a]		75[a]	
1980	78.2	86.1	77.4	83.2
1985	82.1	99.4	80.2	92.6
1990	86.0	112.5	82.7	101.2
1995	89.2	123.6	84.7	107.8
2000	92.0	136.0	86.1	112.5
2005	94.8	145.3	87.0	118.2

a. From observed data.

The results are shown in Table V and the targets are plotted in the trapezoids. Note that the percentages for the proportion of civilians employed are fairly stable; the values increase from 39.8% in 1975 (a year of high unemployment) to 41.8% in 2005.

It should be noted that the gross energy targets shown in Fig. 7 are much lower than those published in recent studies which have ranged from 124 to 225 Quads in 2000 A.D. The lower values resulted from the fact that the rate of growth of both the population and the work force is expected to decline over the next 30 years, a circumstance that has not been given proper weight in studies made to date. The targets also may have to be relocated in the lower part of the trapezoid to allow for the effect of conservation measures. This contingency is discussed in the next section.

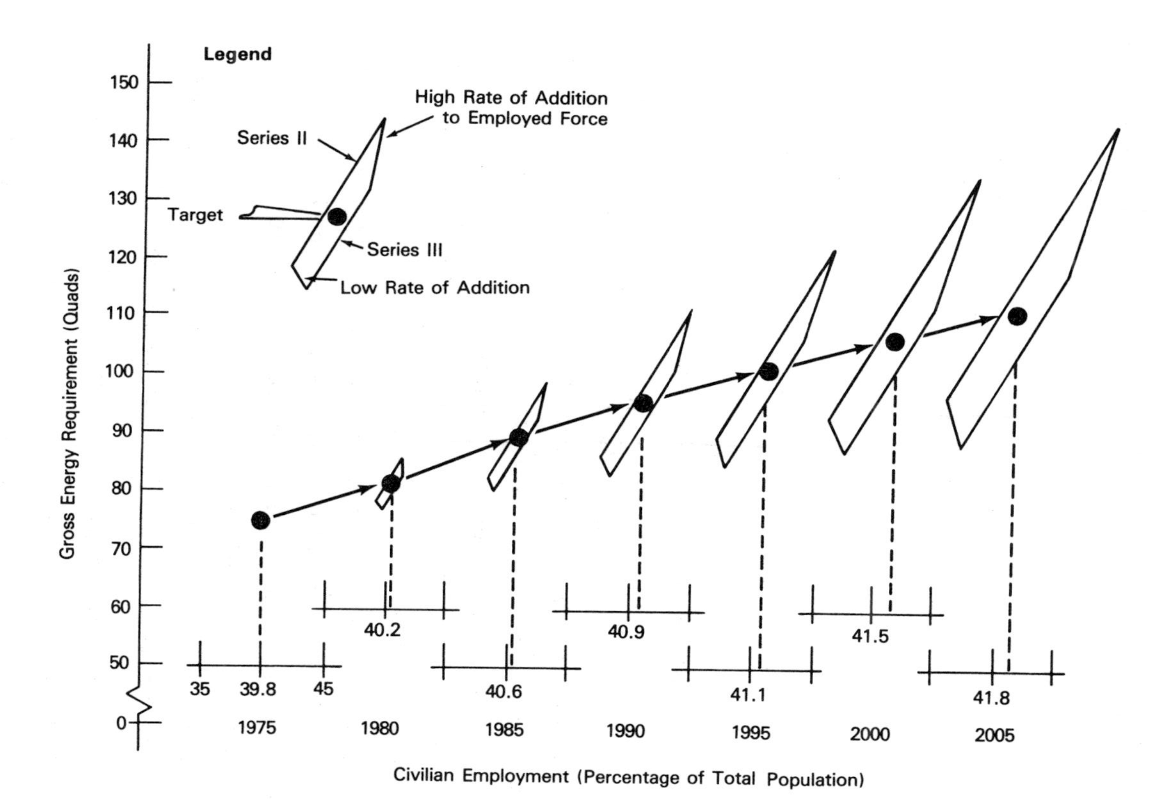

FIGURE 7. Provisional targets to meet energy requirements.

TABLE V. Target Data (Centroids)

Year	*Percentage employed*	*Energy requirement (Quads)*
1975	*39.8*	*75.0*
1980	*40.2*	*81.2*
1985	*40.6*	*88.6*
1990	*40.9*	*95.6*
1995	*41.1*	*101.3*
2000	*41.5*	*106.6*
2005	*41.8*	*111.3*

IV. THE RELATION OF CONSERVATION TO EMPLOYMENT

Let us assume that the objective of conservation is to reduce the rate of growth of energy consumption. The past rates of growth expressed as Quads per year are indicated by the regression coefficients in the following equations:

$$\text{1947 thru 1962 (Ta):} \quad G = 0.9792\ Ta + 30.72 \qquad (3)$$

$$\text{1963 thru 1975 (Tb):} \quad G = 2.230\ Tb + 47.71 \qquad (4)$$

where T subscript represents the number of a given year in the sequence. For example, 1956 is the tenth year in the sequence for equation (3). The gross energy requirement (G) would be $(0.9792)(10) + 30.72 = 40.51$.

Using the targets (centroids) in Fig. 7, the projected rates of growth from 1975 to 2005 are as follows:

$$\text{1975 thru 1990 (Tc):} \quad G = 1.384\ Tc + 74.0 \qquad (5)$$

$$\text{1990 thru 2005 (Td):} \quad G = 1.048\ Td + 94.5 \qquad (6)$$

A moderate reduction in the annual rate of growth of energy consumption of 25% and a greater reduction of 50% for the two periods of time in equations (5) and (6) were computed. The results are plotted in Fig. 8. The top line represents the position of the centroids from Fig. 7. The bottom line shows the position of the targets if a 50% reduction is to be achieved. For example if a 50% reduction is desired, the target for the year 2000 would be about 90 Quads.

The projected reduction could be achieved in several ways. Two are selected to illustrate the relationship between conservation and employment. The first would allow the prevailing rate of energy consumption to continue, namely: 1.55 billion Btu per employee. The reduction in energy consumption could be achieved by dictating a ceiling based on availability of resources, on our national capability to deliver energy (e.g., 90 Quads in the year 2000) or by letting some crisis situation dictate the amount available (e.g., the natural gas shortage last winter). The manpower requirements under this mode of conservation are shown in Fig. 9. For example, if 90 Quads of energy were available in the year 2000 and the rate of consumption per employee is maintained at the prevailing level, then the number of employees would be 95 million. This would represent an employment rate of about 37% of the total population--a reduction of about 10 million people below the targeted level. Obviously 10 million potential workers added to the rolls of the unemployed would not be an acceptable alternative.

The second mode would be to reduce the energy consumption per worker, but at the same time attempt to maintain employment at the targeted level, 105 million employees in the year 2000. The curves for the reduction of individual energy consumption to achieve both a 25% reduction and a 50% reduction in overall energy consumption are shown in Fig. 10. Note that to achieve the desired 50% reduction by the year 2000, the individual employee's consumption would have to be reduced to 1.41 billion Btu. This is a reduction of 10%. Even a modest reduction across

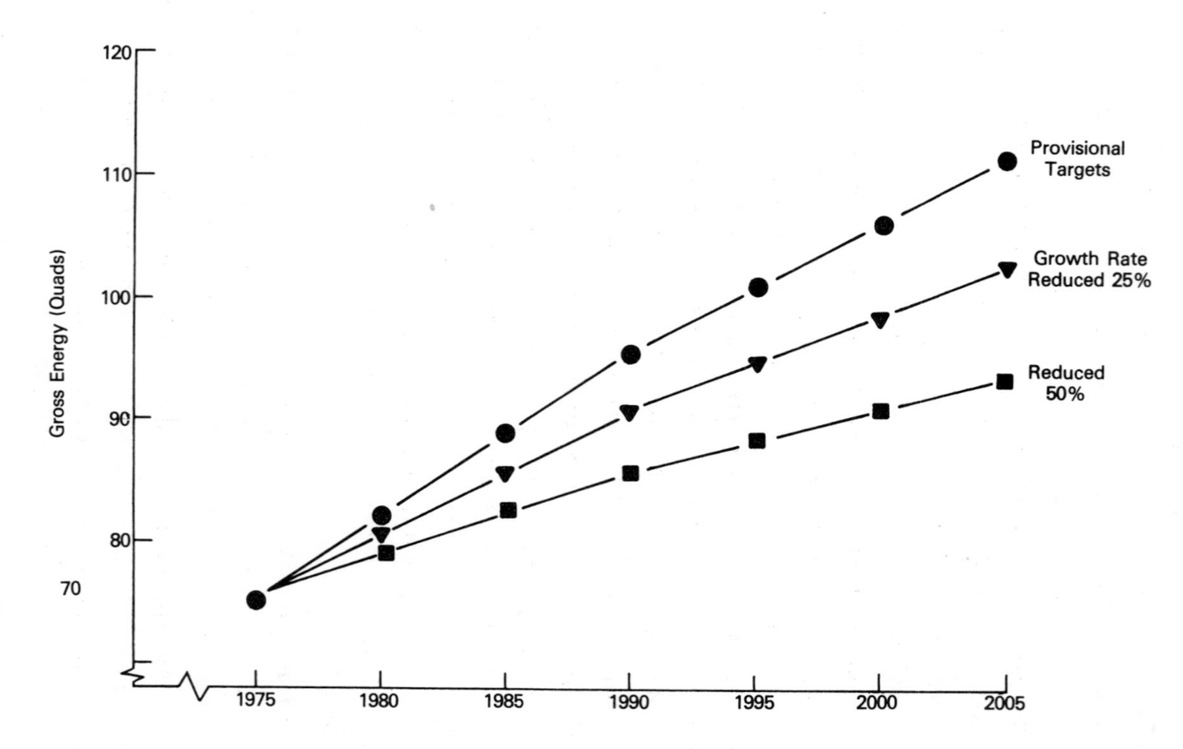

FIGURE 8. Energy requirements with reduced growth rates.

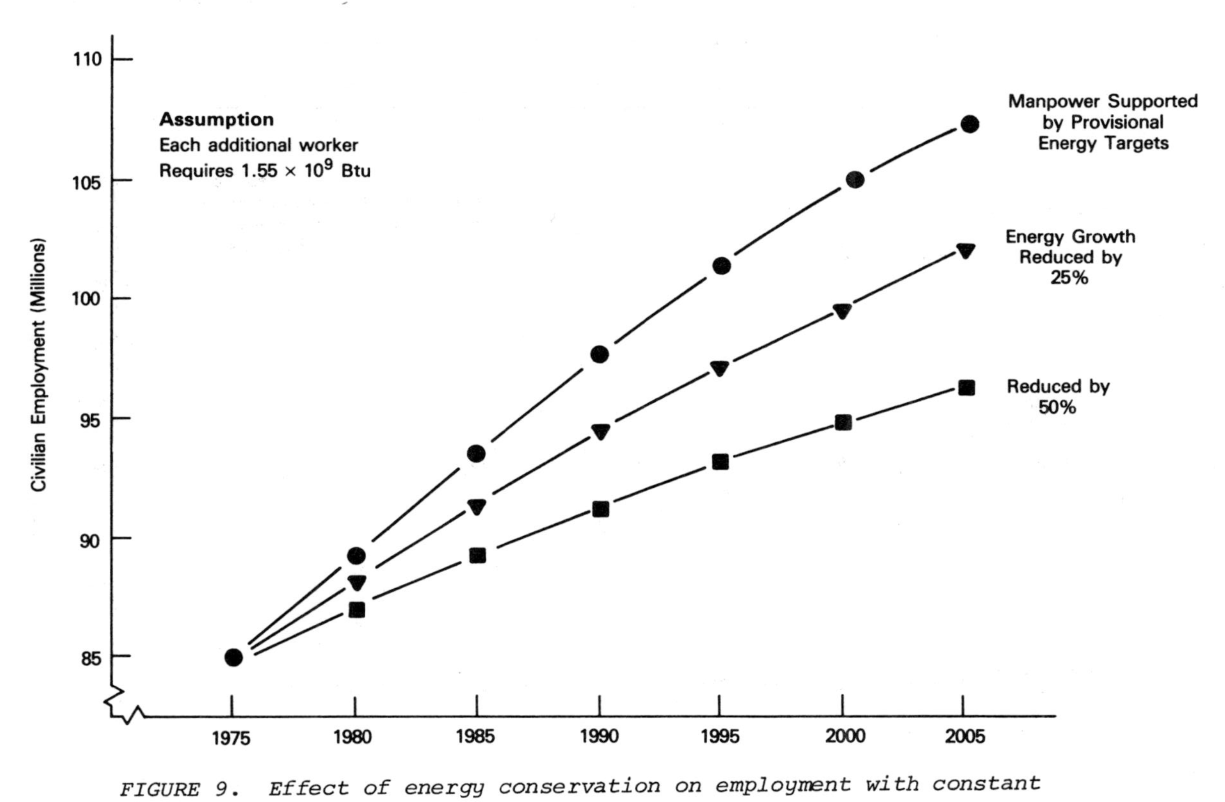

FIGURE 9. Effect of energy conservation on employment with constant usage per employee.

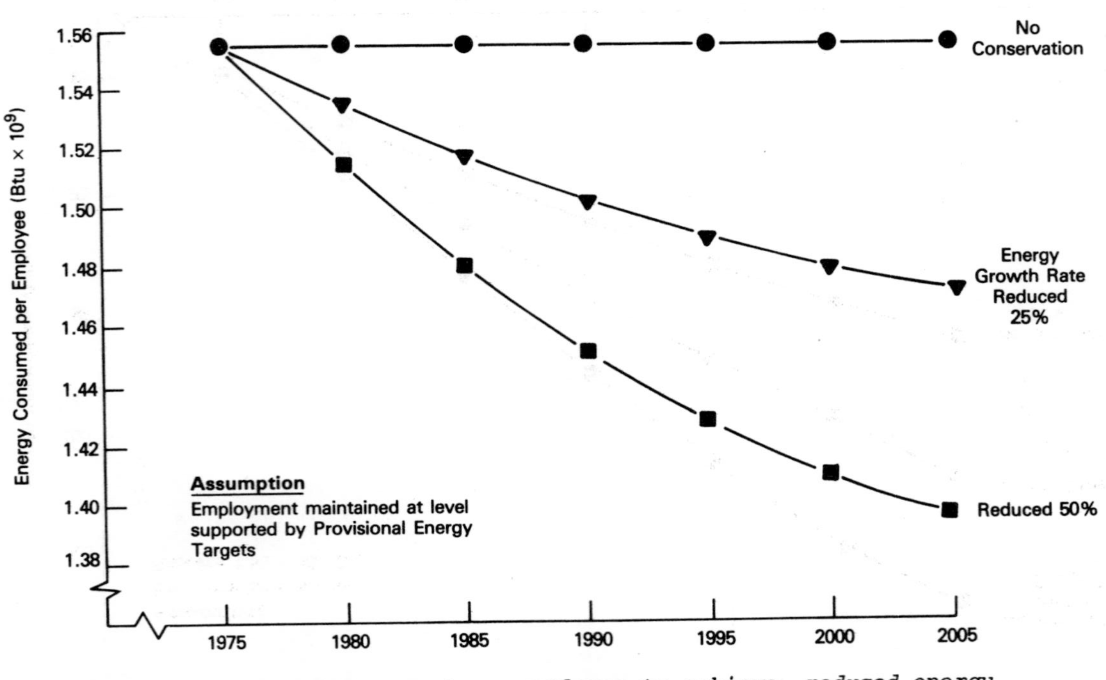

FIGURE 10. Energy consumed per employee to achieve reduced energy growth rates.

the board over the next 23 years will require the progressive application of a variety of conservation measures. Greater use of waste materials must be achieved, and technologies for accomplishing this are discussed in papers by Dr. Brink, Mr. Voss, Dr. Knight, Dr. Lalor, Mr. Arlington, Mr. Greco, and Mr. Varani Less wasteful equipment is also essential along with more efficient industrial processes. Among these alternatives is the cogeneration of electricity and process steam, as discussed in the paper by Dr. Johanson and Dr. Sarkanen.

V. CONCLUSIONS

The data I used in this analysis come from the traditional sources. They probably underestimate the true rate of energy consumption. The discrepancy could be an important consideration when conservation measures are brought into the picture. The data used do not include the energy being derived, or that could be derived, from wastes and renewable organic resources. This information should be included in future compilations and analyses. Mr. Tillman's paper initiates such a discussion.

It seems evident, then, that the requirements approach can make a valuable contribution to energy policy formulation. The relationship between employment and energy is very strong and, if properly used, it could provide reliable estimates of future requirements as the basis for setting policy objectives. This study also reveals that the energy requirements are probably much less than other studies have suggested. The amounts estimated by others for the year 2000 have ranged from 124 to 225 Quads. The rationale used in this study places the requirement for the year 2000 at about 106 Quads with an employment of 41.5% of the total population. This estimate is without the application of conservation measures. A modest reduction of individual energy consumption per worker amounting to 10% by the year 2000 could reduce the requirement to about 90 Quads.

Since the U.S. now consumes about 75 Quads per year, the increment which must be made up is 15 to 30 Quads. Succeeding papers focus on the ways and means for meeting part of this increment from renewable resources. As these papers demonstrate, from 10 to 40% of that increment could be obtained from renewable organic resources.

REFERENCES

1. Population Estimates and Projections. Series P-25, No. 541, Department of Commerce, Feb. 1975.
2. United States Energy Through the Year 2000 (Revised). U.S. Bureau of Mines, Department of the Interior, Dec. 1975.
3. Economic Report of the President. Transmitted to Congress, Jan. 1977.

UNCOUNTED ENERGY:

THE PRESENT CONTRIBUTION OF RENEWABLE RESOURCES*

David A. Tillman

Materials Associates
Washington, D.C.

I. INTRODUCTION

In November, 1976, Durkee Manufacturing Co. of Pine River, Minnesota, installed a wood-fired boiler to supply 100% of its energy needs. They joined a large and ever increasing number of firms switching to nonfossil organic fuels: renewable resources. Federal statistics fail to account for the use of these fuels. Yet such renewable resources are gaining increasing recognition among industries and home owners. The current energy situation in the U.S. makes this recognition essential.

A. The Need for Analysis

As Dr. Mottley demonstrated in the previous paper, U.S. energy demand is rising inexorably as jobs are created for new

---

* This research was partially supported by Federal Energy Administration Contract No. P-03-77-4426-0.

labor force entrants. At the same time, domestic supplies of oil and gas are declining. Expensive oil imports from less than stable sources, now used to make up the difference, cost this country $25 billion in 1975 and over $30 billion in 1976. Limited reserves and resources of uranium cloud the nuclear power picture. Only coal among the primary domestic energy sources is showing real prospects of long-term growth. Compounding the problem is the lack of a clear-cut long-term energy policy coordinating all aspects of the problem.

History is replete with examples of civilizations that failed to solve the problem. The ancient Babylonians discovered deposits of bitumen: the first recorded surface expressions of Middle East oil. They used this fossil fuel for space heating and metals smelting. When the bitumen ran out, the civilization disappeared. It cannot be argued that the U.S. is in such shape. At the same time, the illustration is instructive. This country needs to look to alternative fuels for increasing amounts of energy.

The CORRIM Report of the National Academy of Sciences observes, generally, that oil and gas supply is a long-term problem that can be met, at least partially, by the use of renewable resources. That report considers the combined economic and environmental cost/benefit ratio of using renewable resources to be quite favorable [1]. Thus an assessment of the present contribution now is critical.

B. The Nature of This Analysis

In order to examine the contribution of nonfossil organic energy resources, this paper makes a cursory survey of the historical antecedents of renewable energy resource use. It proceeds to a detailed examination of the present contribution of silvicultural, agricultural, and urban residues and fuels. Finally, it documents trends and forces associated with the use of renewable resources. From these data conclusions can be drawn.

1. Fuel Materials Included

This paper considers a broad range of materials as nonfossil organic energy sources. The following broad categories and specific subcategories are included: (1) Silvicultural materials --cord wood, charcoal, wood chips or "hogged fuel," bark, sawdust, and spent pulping liquors; (2) agricultural materials--bagasse, cotton gin trash, wheat hulls and other grain residues, and animal manure; and (3) urban wastes--general and specific manufacturing wastes (other than wood processing wastes), municipal solid waste, and methane-rich gas from landfills and sewage treatment plants.

2. Energy Uses Considered

For the purposes of this report, an energy use includes the following:

a. providing process heat for industrial or electricity generating purposes,

b. providing space heating or cooking, in industrial, commercial, or residential applications,

c. providing a carbon source for the reduction of metals (significant because the only alternatives found to date have been fossil fuels).

II. HISTORICAL SURVEY

Wood was probably mankind's first fuel, and its contributions to the development of civilization cannot be overemphasized. Herman Kahn argues, successfully:

> When primitive man learned to make fire, he had discovered controllable energy, which then

became a "servant" destined to perform an endless series of "miracles," beginning perhaps with simple cooking. This discovery may have been the single most vital factor which allowed mankind to develop modern civilization [2].

A. The Development of Civilization

The development of civilization can be traced through the mining, refining, and use of metals. The use of fuels follows along a parallel, related path.

Copper was among the first base metals to be relatively widely used in society. It was first smelted in small furnaces around 3500 BC. Those furnaces were of a natural draft type and fueled with wood [3]. Fire refining, and repetitive fire refining, were practiced in Mesapotamia as early as 2000 BC [3]. During the Greco-Roman Era, the island of Cyprus rose to prominence in the world of trade. Its copper mines were worked; and bronze swords, armor and lance tips were made for export. Cyprian metallurgy had reached a high degree of perfection, and the vast forests of that island fueled the copper smelters, and hence the export economy [3]. The copper mines of Sweden, at Stora Kopperberg, were used by the royal family to finance wars from 1288 to 1650. Copper was extracted by lighting wood fires at the mine face and then using hand tools to recover the ore. Smelting was performed in vertical shaft furnaces fueled by both wood and charcoal. Stora Kopperberg was considered a most valuable possession by the crown [3].

The need for and use of metals continued to expand through the centuries. By the 14th and 15th centuries, Europe's metallurgy industry faced a serious crisis. Deforestation resulted from wood fuel and charcoal production [3]. In some areas, constricted metals production resulted. In the Vizcaya district of Spain, however, a law requiring six trees to be planted for every tree removed sustained the supply of this renewable resource [3]. In England, severe deforestation forced another solution: one that reduced and ultimately eliminated the use of wood as an industrial fuel in that country. Abraham Darby produced coke from coal and substituted it in a blast furnace in Shropshire, in 1709 [4]. By 1750 the iron industry had moved from such regions as the Forest of Dean and Weald to the Birmingham area, as coal and coke had displaced relatively scarce wood [3].

B. The Economic Growth of the U.S.

Wood was the first and most obvious fuel for this nation. The first wood stoves for New England were cast in a Saugus, Massachusetts foundry in 1647. The Franklin stove was invented in 1742, and it appeared in New England within a decade. Soapstone stoves, which held the heat of a fire long after the flames and coals had died, were made at that time in New Hampshire [5]. In addition to being a residential fuel, wood served the industry of a new nation. By 1838 some 1800 steam engines existed--fueled by wood [4]. Little coal was used for industry at that time.

In 1850, 90% of all U.S. energy came from wood, and 100 million cords of that fuel were used annually [6]. Although most of this fuel was used in households, industry provided a strong market. More than half of the iron made was smelted with charcoal, and consumption levels had reached 70 million bushels [6].

In 1880, 150 million cords of wood were used totally. Some 1.5 million cords were being turned into 86 million bushels of charcoal for metal smelting. The railroads remained almost totally dependent on wood and were consuming six million cords annually [6]. Ironically, in that year when wood contributed

2.89×10^{15} Btu to the economy [6], coal surpassed it as this country's primary fuel.

Wood, as a fuel, began to decline after that decade. Even in decline, however, wood technology continued to advance. When the Wier Stove Co. of Taunton, Massachusetts, brought out the advanced "Glenwood" line of residential heating and cooking equipment in 1900, a boom occurred. In the year 1920 that firm built a new 150,000 sq. ft. factory, employing 850 workers to turn out the products [5]. Sawdust burners were introduced and gained commercial acceptance toward the end of the 1920s, particularly in the Pacific Northwest [7]. Their average efficiency of 40.6% was most advanced [7]. In 1950, the last year for official published government data concerning wood combustion, some 4.3 million homes still used this fuel. In that year it contributed 1.19×10^{15} Btu to U.S. energy supply [6].

Crop and animal residues have also been associated with U.S. energy development. Methane production by anaerobic digestion gained much of the attention. In 1774, Benjamin Franklin wrote to his friend Joseph Priestly concerning methane release from swamps. Then, as the U.S. Civil War was coming to a close, the French scientist Pasteur established that methane production results from the action of living organisms. A decade later, in 1875, Popoff demonstrated that cellulose could be decomposed with living organisms for the specific and intended production of methane [8]. The first U.S. patents for processes to produce methane by anaerobic digestion were issued in 1904 [8]. But the principal research resulting in a workable system was carried on in India. At the Agricultural Research Institute in New Delhi, the Gobar Gasifier emerged in 1939 [10].

Fermentation of crop residues into ethanol and methanol was also well established technically. The age old practice of making "white lightning" illustrates the widespread nature of fermentation practices. In the 1920s, the U.S. Department of Agriculture began publishing bulletins on the use of fermented alcohols in farm equipment, as fuel. During the depression and World War II, ethanol was a common fuel in the midwest. In the same era, Europe was consuming 500,000 tons of plant matter for conversion into fuel alcohol [11]. Radical declines in the price of petroleum, however, virtually eliminated this practice.

III. THE PRESENT CONTRIBUTION OF NONFOSSIL ORGANICS

During this decade, the use of renewable resources rebounded. Led by a resurgence in the use of wood as fuel, they now make a substantial contribution to the economy.

A. The Use of Silvicultural Materials

To analyze the increased use of wood, this presentation analyzes its contribution to the following economic sectors: (1) pulp and paper, (2) lumber and plywood, (3) metals refining, (4) other industries, and (5) residential applications. Charcoal is considered in the residential section.

1. Pulp and Paper Mills

In an effort to dispose of wastes and simultaneously conserve on premium fossil fuels, the pulp and paper industry has turned increasingly to wood residues: hogged fuel, bark, and spent liquor. Direct combustion and fluidized bed combustion have both been used to convert these materials into useful energy. Table I documents the energy from renewable resources used by the pulp and paper industry. It is rapidly approaching one quadrillion Btu (or quad) [12].

During this period of time, renewable resources increased their share of the total energy supply in the pulp and paper industry from 36 to 44.3%. In some regions, they became the principal fuel supply. Table II presents this contribution estimate from 1976.

TABLE I. Use of Residues as Fuel in the Pulp and Paper Industry[a]

Type of residue	Year		
	1971	1973	1976[b]
Wood chips	10.0	37.0	83.0
Bark	113.0	114.0	98.0
Spent liquor	667.0	681.0	802.0
Total	790.0	832.0	982.0

a. In 10^{12} Btu.

b. Based on first six months utilization rates.

TABLE II. Use of Residues as Fuel in the Pulp and Paper Industry as a Percentage of Total Energy Consumption (by region)

Region	Residue Type			
	Wood chips	Bark	Spent liquor	Total
New England	2.3	3.9	23.2	29.4
Middle Atlantic	-	3.0	15.3	18.3
North Central	0.9	1.9	12.9	15.7
South Atlantic	3.0	6.3	43.2	52.5
South Central	3.3	5.9	41.2	50.4
Mountain and Pacific	9.0	0.3	42.9	52.2
National average	3.7	4.4	36.2	44.3

This trend is increasing in the pulp and paper industry: the largest manufacturing consumer of fuel oil. The Union Camp Corp. expansion in Franklin, West Virginia, will boost pulp output by 13%. The residue fired energy recovery system will supply 1.5×10^{12} Btu and reduce expected purchased energy needs by 20% [13]. Groveton Papers recently added the municipal waste of Northumberland, New Hampshire, to its bark and hogged fuel system, saving an additional 10,000 bbl of oil per year [14]. The 1000 ton per day pulp mill being built by Parsons and Whittemore in Claiborne, Alabama, will have a vast energy recovery-pollution control system [15]. The 5% annual growth rate for use of renewable energy resources, experienced by the pulp and paper industry since 1971, can be expected to continue for the foreseeable future.

2. Lumber and Plywood Mills

Sawmills and plywood mills also gain much of their energy from wood residues. Their annual consumption is estimated to be 70×10^{12} Btu [16]. The distribution is evenly split: 35×10^{12} Btu for sawmills and 35×10^{12} Btu for plywood mills. For the sawmill industry, 35×10^{12} Btu represents 20 to 40% of the total energy requirement. For the plywood industry, it represents 50% of the annual requirement

Significantly, while the pulp and paper industry is oriented almost totally toward direct combustion systems, pyrolytic conversion may emerge in the sawmill industry. Tech Air Corp. operates a 50 ton per day unit at Cordele, Georgia. It produces a range of solid liquid and gaseous fuels. Forest Fuels, Inc., New Hampshire, has built a pyrolysis gasifier for use in the kiln-drying phase of lumber production and has completed its first installation. The Kiersarge Reel Co. of Maine produces some 40×10^{9} Btu/year with this gasifier, eliminating the need to purchase 6600 barrels of #2 heating oil annually [17]. American Fyr Feeder has also entered the wood gasification field.

This increased diversity in energy approach may spur increases in the use of wood residue.

3. Metallurgical Industries

Refiners of ferronickel, ferrosilicon, and copper use varying amounts of wood in their production processes. Wood wastes are used extensively in ferronickel production. The Riddle Mt., Oregon, deposit (being worked by Hanna Mining Co.) supplies 10% of U.S. ferronickel requirements. That company uses a process developed by the U.S. Bureau of Mines [18]. Hanna Mining uses wood chips and sawdust as a heat source and as a reductant. The nickeliferous ore is dried from 21 to 3-5% water in chip fired rotary driers. Sawdust is used as a prereductant in the calcining step. Then chips are used to reduce ferrosilicon, since ferrosilicon is the reductant for ferronickel [19].

The 18 operating ferrosilicon plants in the U.S. use wood chips as a carbon source. Their high electrical resistivity is an asset. Further, the hogged fuel adds bulk and stability to the electric furnace charge and helps prevent bridging.

To obtain data on the 1976 consumption of hogged fuel by ferroalloy producers, the producers of ferrosilicon and ferronickel were surveyed in December, 1976. Table III presents the results of that survey, which reflected production at all operating ferrosilicon and ferronickel plants in the U.S. The 100% response offering the 11.666×10^{12} Btu provides the first precise estimate of their renewable resource utilization.

Two copper smelters also utilize wood, employing the ancient poling process handed down since at least the 14th century to reduce the oxygen content of copper from 0.9 to 0.2%. Most copper smelting and refining operations now use natural gas to accomplish deoxidation of the molten red metal. White Pine and the ASARCO-Takoma, Washington, smelter still uses wood poles. It takes 15 poles, each 1 ft in diameter and 30 ft long, to deoxidize 240 tons of copper [20]. At current rates of production

TABLE III. Use of Hogged Fuel in Ferroalloy Production, 1976

Firm classification (by tons of wood chips used)			*Number of firms*	*Total wood chip fuel consumption (in 1 × 10^9 Btu)*
0	<	*1,000*	*4*	*17*
1,000	<	*10,000*	*4*	*362*
10,000	<	*25,000*	*3*	*847*
25,000	<	*50,000*	*3*	*1,634*
50,000	<	*100,000*	*2*	*1,836*
	>	*100,000*	*3*	*6,970*
Total responses			*19*	*11,666*

at White Pine and ASARCO, some 300 × 10^9 Btu of wood are used in the refining of this metal.

Total renewable resource utilization in metals refining for 1976 can therefore be estimated at 12.0 × 10^{12} Btu.

4. Other Industries

Numerous other firms use wood and wood waste as fuel. These include millwork companies, furniture plants, apparel companies, the Eugene Water and Electric Board, the University of Oregon, and a host of others. One sugar refinery in Hawaii uses hogged fuel. The list is legion [21-24].

Adding up examples case by case becomes a tedious and self-defeating process. Estimates made by Oregon State University engineers indicate that some 1000 wood-fired boilers exist in the western U.S., and that 15% are in industries other than silvicultural or metallurgical concerns [25]. A survey of 172 installations of Fyr-Feeder wood-burning boilers indicates that 78.5% are located in pulp and paper, lumber, or plywood concerns;

16.3% were installed in furniture and millwork plants, and 5.2% were located in such other places as the U.S. Public Health Service in Atlanta, Georgia.

One can assume that boilers installed outside the silvicultural industries will be somewhat smaller than those in the pulp and paper, lumber, plywood, and veneer mills. If one assumes that wood consumption outside the silvicultural community will be only 10% of consumption inside that industry grouping, then an estimate of 105×10^{12} Btu can be used for this group of organizations.

5. Residential Uses and Charcoal Production

In 1970, some 800,000 houses were heated with wood, consuming an estimated 200×10^{12} Btu/yr [26]. Estimates for 1972, published in 1973, place fuel wood consumption at 300×10^{12} Btu/yr [24]. Since that time the sale of residential wood-fired equipment and cordwood has literally zoomed. Cordwood in the Northeast has sold for an average of $75/cord, or $3.75/MM Btu.

To estimate with some degree of confidence the present consumption of fuel wood, a questionnaire survey of 36 wood-burning equipment manufacturers was conducted. The sample consisted of the U.S. Forest Products Laboratory's partial list of companies in the field. Nine firms responded, seven filling out the questionnaire and two with comments only. Table IV presents the responses. Of these sales, the companies estimated that 66.7% of the equipment was sold for supplementary heating and cooking, and 17% for primary heating and cooking.

The additional comments were as follows:

> Our products were released in August of 1975. Our 1976 sales were up 175% over '75. We are expecting a 400% increase in '77 sales over '76.

> I am afraid that we cannot fill out your questionnaire, due to the fact that we have been

TABLE IV. Estimated Annual Increase in Equipment Sales of Residential Wood-Fired Equipment

Percent annual increase in sales since 1972	*Number of firms responding to questionnaire*
Negative increase	*0*
0%	*0*
0 < 25%	*2*
25 < 50%	*2*
50 < 75%	*0*
75 < 100%	*2*
100 < 200%	*0*
200 < 300%	*1*
> 300%	*0*
Total	*7*

> manufacturing stoves here for only the last year. During that period we have made and/or marketed about 3,500 stoves in four different sizes.

Finally it has been noted that wood-fired systems have been hot items at auctions [5] and that equipment dealers have had a field day in New York and Pennsylvania [27].

Based upon the above data, principally on the dramatic increase in sales of wood utilization equipment, one can estimate a doubling in the rate of fuel wood utilization in residential applications since 1970, or a 50% increase since 1972. While these data defy precise and rigorous statistical analysis, a total consumption rate of 400×10^{12} Btu appears reasonable.

In addition to the sales of wood-fired equipment, it has been observed that some 720,000 tons of charcoal (85% made from wood) were produced in 1976. This represents an additional

15×10^{12} Btu of wood-based fuel used in the home. This is a 10.1% increase over 1974 production [28].

6. Composite Picture of Wood and Wood Residue Utilization as Fuel

Table V presents the total use of wood and wood residue as fuel. Total consumption in 1976 was 1.58 quads.

B. The Use of Agricultural Materials

Agricultural materials can be divided into two kinds of waste: crop waste and animal waste. Both types of material are used for energy.

1. Crop Waste Utilization

To date, bagasse and cotton gin trash have emerged as the principal crop waste energy sources. Bagasse, the residue from cane sugar refining, has been used as a fuel for decades. In 1976 3.7 million tons of bagasse were used as fuel, supplying the economy with 25×10^{12} Btu [29]. This supplies 63% of that industry's energy needs. (See paper by Mr. Arlington.)

Cotton gin trash utilization is only now emerging as an energy source. Two installations have been completed to date. The Kiech-Shauver cotton gin in Monette, Arkansas, installed the first energy-recovering incinerator in that industry. That firm gins some 8200 bales of cotton per year, recovering an average of 115,549 Btu/bale. It is estimated that total annual energy recovery was 0.95×10^{9} Btu in 1976 [30]. (See paper by Dr. Lalor.)

2. Animal Waste Utilization

Methane-rich gas can be produced from manures of all types: hog manure, chicken excrement, and bovine dross. These wastes may be generated either in small concentrations (i.e., on dairy farms) or in large centralized operations (i.e., feedlots).

TABLE V. Wood Utilization as Fuel for 1976

User Group	*Wood and wood residue utilization (in 10^{12} Btu)*
Pulp and paper	*982*
Sawmills, plywood mills, and veneer mills	*70*
Metallurgical industries	*12*
Other industries	*105*
Residential	*400*
Charcoal	*15*
Total	*1584*

a. Small-Scale Applications. Dairy farms, chicken farms, and hog farms provide locations for small-scale applications of anaerobic digestors. In these locations, Gobar-type gasifiers can be utilized. Sharon and James Whitehurst, for example, imported Gobar technology directly. Originally they built a 225 ft^3 gasifier, producing 80×10^6 Btu/yr on their Vermont farm. Since that time they have built a 4500 ft^3 gasifier producing 1.4×10^9 Btu/yr [31]. At the 1973 Pennsylvania Agricultural Progress Days at Hershey, Pennsylvania, two similar gasifiers were exhibited, receiving much attention and interest [32]. Since those two events, numerous articles have been published by John Fry of the New Alchemy Institute and by others on how to build such systems. From time to time local newspapers publish articles on this farmer or that farmer building and using small-scale gas-producing systems.

b. Large-Scale Applications. The use of anaerobic digestion on cattle feedlots has been proposed for many years. This process capitalizes on the moisture in manure and also preserves nutrients for fertilizer applications of the sludge. It appears to have a competitive advantage when compared to pyrolysis or other systems favoring a dry feed [33]. To date, no systems have been built. One has been proposed for Monfort of Colorado's 100,000 head feedlot outside Greeley, Colorado. As designed it will produce 2×10^{12} Btu annually [34]. Three others are in design or construction phases. (See paper by Mr. Varani.)

3. Aggregate Data

In total, agricultural residues supply over 25×10^{12} Btu/yr. Virtually all of this comes from bagasse, but an immediate potential exists in the use of cotton gin trash and manure.

C. Urban Waste Utilization

Urban waste consists of manufacturing residues, municipal solid waste, and methane from landfills and sewage sludge. The generation of residues, at this stage in the materials cycle, is less than the generation of waste materials in the extractive industries. Counterbalancing that influence, however, is the degree of concentration and proximity to markets for energy fuels.

1. Industrial Residues

Organic industrial residues come in a variety of forms: meat packing waste, film waste, and general manufacturing waste. Table VI presents a representative selection of these materials. One conservative survey places the total generation of nonwood commercial and industrial waste generation rate at 44.0 million tons per year: equal to 600×10^{12} Btu/yr [35]. (It should be noted that wood residues from furniture and millwork manufacturing, treated earlier in this paper, have been eliminated from this discussion to obviate problems associated with double counting.)

TABLE VI. The Production of Selected Industrial Residues

Industry	*MM tons of residue generated per year*	
	1965	*1975*
Food	*5.3*	*7.0*
Meat	*0.8*	*1.2*
Textile mill products	*0.9*	*1.1*
Apparel and related products	*0.3*	*0.5*
Printing and publishing	*1.2*	*1.6*
Chemicals	*1.3*	*2.4*
Rubber	*1.5*	*1.9*
Tanning	*0.3*	*0.3*
Machinery	*4.4*	*6.9*
Supermarkets	*10.2*	*9.4*

Source: National Commission on Materials Policy, Material Needs and the Environment Today and Tomorrow, U.S. Government Printing Office, June, 1973.

An increasing number of companies are using their own waste as an energy source. Examples of such utilization include installations at the following companies: General Motors [36], Eastman Kodak [37], John Deere [38], Goodyear Tire [39], and Xerox Corp. [40]. John Deere is emerging as a leader in this area with installations at their Horicon, Wisconsin, works and their Dubuque, Iowa, plant, and one to be installed at the new Waterloo, Iowa operations [41]. The magnitude of energy production at these installations, shown in Table VII, is 2×10^{12} Btu/yr.

In addition, a cursory literature search indicates that a major Texas paper and poly film converter of packaging materials pyrolyzes its waste for energy recovery [42]. A large producer of cellulosic and synthetic fibers does the same [42]. Instant coffee manufacturers burn coffee grounds to produce steam for fresh coffee processing [43]. One such firm operates five 40,000 lb/hr steam boilers for recovering energy from its residues [44]. Peanut hulls are used in peanut butter production also [44].

It can be estimated that 15% of these nonwood processing wastes are converted into energy each year, yielding a total energy production of 90×10^{12} Btu/yr.

TABLE VII. Identified Companies Producing Energy from Industrial Waste

Company	*Estimated energy recovered/year (in 10^9 Btu)*
John Deere Horicon Works, Horicon, Wisconsin	*15*
John Deere, Dubuque, Iowa	*15*
Xerox, Columbus, Ohio	*46*
General Motors, Truck and Coach Div. Detroit, Michigan	*910*
Eastman-Kodak, Rochester, New York	*650*
Goodyear Tire, Jackson, Michigan	*340*
Total	*1976*

2. *Municipal Solid Waste*

Numerous processes now exist and are being deployed to convert municipal solid waste into useful energy. These processes include incineration, co-combustion of waste and coal, pyrolysis, and anaerobic digestion. Table VIII presents a list of plants, currently on-line, converting 41.2×10^{12} Btu of energy contained in municipal waste into useful energy.

TABLE VIII. Present Plants Recovering Energy from MSW

Present Plants	*Size tons/day*	*Annual energy value (1×10^{12} Btu)*
Ames, Iowa	*200*	*0.70*
So. Charleston, West Virginia	*200*	*0.70*
San Diego, California	*200*	*0.70*
St. Louis, Missouri	*325*	*1.14*
Milwaukee, Wisconsin	*1200*	*4.20*
Baltimore, Maryland	*1000*	*3.50*
Baltimore County, Maryland	*1200*	*4.20*
Chicago, Illinois	*1600*	*5.61*
Chicago, Illinois	*2000*	*7.00*
Nashville, Tennessee	*720*	*2.52*
Harrisburg, Pennsylvania	*720*	*2.52*
Saugus, Massachusetts	*1200*	*4.20*
Norfolk, Virginia	*360*	*1.26*
Braintree, Massachusetts	*240*	*0.84*
Ft. Wayne, Indiana	*300*	*1.05*
Bridgewater, Massachusetts	*300*	*1.05*
Total	*11,765*	*41.19*

3. Landfill and Sewage Sludge Gases

Methane-rich gases containing 300-700 Btu/scf are produced by anaerobic digestion of municipal waste in landfills and of sewage sludge. In the case of landfill gas, utilities are showing increased interest. Sewage sludge gas normally is employed to provide energy to run the treatment plant itself.

Three methane-rich gas recovery projects are now underway to recover useful energy from landfills. The 2 MM cf/day raw gas recovery project at Palos Verdes, California, was the pioneer project. There raw gas is upgraded to 1 MM cf/day of methane and sold to Southern California Gas Co.[45]. The city of Mountainview, California, and Pacific Gas and Electric Co. have a similar arrangement in northern California [46]. Finally, the Los Angeles County Sheldon-Arleta landfill area has been drilled to yield 1000 cfm of 500 Btu/cf gas. This gas is sold to the Valley Steam Plant [47]. None of the landfill projects offer a great supply of energy. Palos Verdes yield is estimated at 0.35×10^{12} Btu/yr. The Mountain View yield is about the same. Sheldon-Arleta is somewhat larger, offering an energy production rate of 0.5×10^{12} Btu/yr. The total annual production from existing projects then is about 1.2×10^{12} Btu/yr.

Sewage provides far more energy than landfill gas to the U.S. economy. The gases resulting from anaerobic digestion of sewage are used to heat sewage treatment facilities and generate internally consumed electricity. Outside sales of methane-rich gas are rare. In Los Angeles County, the Hyperion Sewage Treatment Plant does sell surplus gas to the Department of Water and Power's Scattergood Steam Plant [47]. There are some 3219 sewage treatment plants in the U.S. with anaerobic digestion systems. Some 51% handle less than one million gallons per day while 30% handle 1.5 MGD and 19% handle more than 5 MGD of sewage [48]. Based upon standard design data of 1 ton of active solids per MGD of flow, and 12 scf of 566 Btu/cf gas per pound of solids, it is estimated that these plants are producing some 47×10^{12} Btu/yr.

4. Total Consumption of Urban Residues

Urban nonwood residues in total supply the economy with 178.8×10^{12} Btu annually. Of this, 90×60^{12} Btu comes from industrial waste, 41.2×10^{12} Btu comes from municipal solid waste, 1.2×10^{12} Btu comes from landfill gas, and 46.4×10^{12} Btu comes from sewage treatment systems.

D. Total Current Contribution of Nonfossil Organics

Nonfossil organic fuels, as renewable resources, contribute a significant amount of energy to the U.S. economy. At present, they contribute in excess of 1.8 quads. Table IX presents that total while Table X comparies it to other energy sources. Nonfossil organics presently contribute as much energy as nuclear power. They contribute 75% of the amount of energy expected from the Alaska Pipeline, and 56% of the amount of energy supplied by hydro electric generating stations. They contribute 2.5% of the nation's total energy supply.

IV. FORCES ASSOCIATED WITH USING NONFOSSIL ORGANICS

Numerous factors explain why nonfossil organic materials are increasing their contribution to U.S. energy supply. In order to develop a clear picture of these forces, certain trends must be isolated and identified. From there, analysis can proceed.

A. Trends in Utilization

Silvicultural materials, the dominant fuel material, have increased their contribution from 1.1 quads to 1.58 quads, a growth rate of 5.1%. These are the base materials. It is

TABLE IX. The Present Contribution of Organic Renewable Resources to U.S. Energy Supply

Type of material/user community	Present contribution (in 10^{12} Btu)
Silvicultural materials	1584
Pulp and paper mills	982
Lumber, plywood, and veneer mills	70
Metallurgical concerns	12
Other industries	105
Residential applications	400
Charcoal	15
Agricultural materials	25
Bagasse	25
Cotton gin trash	neg.
Manure	n/a
Urban materials	179
Industrial (nonwood) waste	90
Municipal solid waste	41
Methane rich gases	48
Total	1788

TABLE X. Comparison of Organic Renewable Resource Fuels to Other Energy Sources

Energy source	Present contribution (in 10^{15} Btu)
Nonfossil organics	1.8
Nuclear power	1.8
Hydroelectric power	3.2
Geothermal	neg.
Alaska oil pipeline (expected)	2.4

significant that all nonfossil organic materials are increasing their energy contribution.

The vast majority of the material consumed traditionally has been burned to supply some energy. These materials --wood, spent liquor and bagasse--are produced and consumed by the same firm. An analysis of trends can focus on such materials. Fig. 1 presents the trends in boiler capacity installations for the years 1963-1975. The numbers presented are the percentage of total installed industrial boiler capacity designed to be fired by wood, spent pulping liquor and bagasse. In 1963, 11.9% of the industrial boiler capacity installed was designed for those three nonfossil organics. That percentage fell to 1.8% in 1967. Subsequently it rose 14.6% in 1975. The 1963-1967 trend is described by the formula $y = 11.82 - 2.28x$, where 1963 is the base year. The 1967-1975 trend is described by the formula $y = 4.24 + 1.3x$, where 1967 is the base year.

The downward trends in boiler installations represents the tail end of a nearly century long trend of declining usage of renewable resources. The upward trend represents the fundamental support for the resurgence of these materials.

B. Reasons for Increasing Utilization of Renewable Resources

Two basic forces can be considered as driving mechanisms in the trend toward increased use of organic renewable resources: energy considerations and environmental regulation.

1. *Energy Considerations*

Energy considerations of price and availability must both be considered. Rising prices could make these fuel forms increasingly attractive. An absolute shortage of fuels could drive companies to use the less concentrated renewable resources.

Oil price appears to have played a significant role in the resurgence of renewable resources. For the years 1967-1973, the equation $y = 8.168x - 17.91$ describes the short-term trend

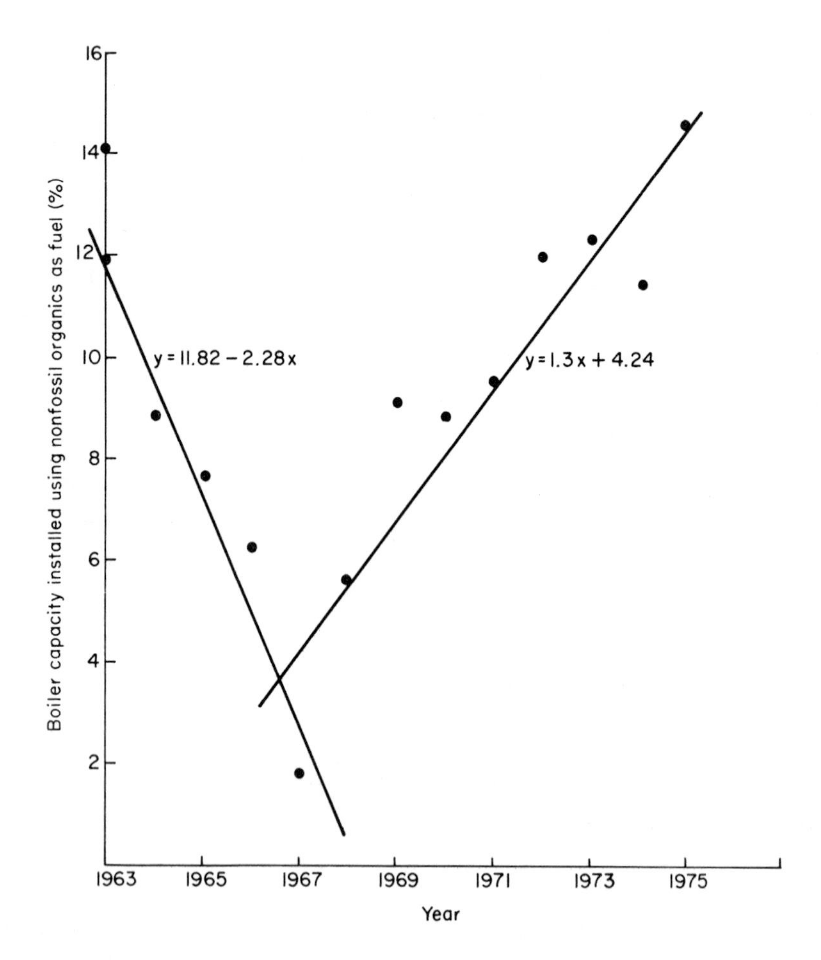

FIGURE 1. Trends in nonfossil organic boiler capacity installation.

depicted in Fig. 2. Three year moving-average values were employed to describe boiler installation trends in order to reduce the impact of economic variations (i.e., the 1971 recession) on short-term trend description. With an r value of 0.9317, and an r^2 value of 0.868, the trend can be considered significant.

The trend disintegrates after 1973, as the 1974 value shows. An oil embargo, and subsequent radical oil price increases destroyed it. (If the trend had held up, 37.14% of all industrial boilers installed in 1974 would have been residue fired.) Subsequent to the oil embargo, a severe recession, a natural gas shortage in the East, and a drought in the West may

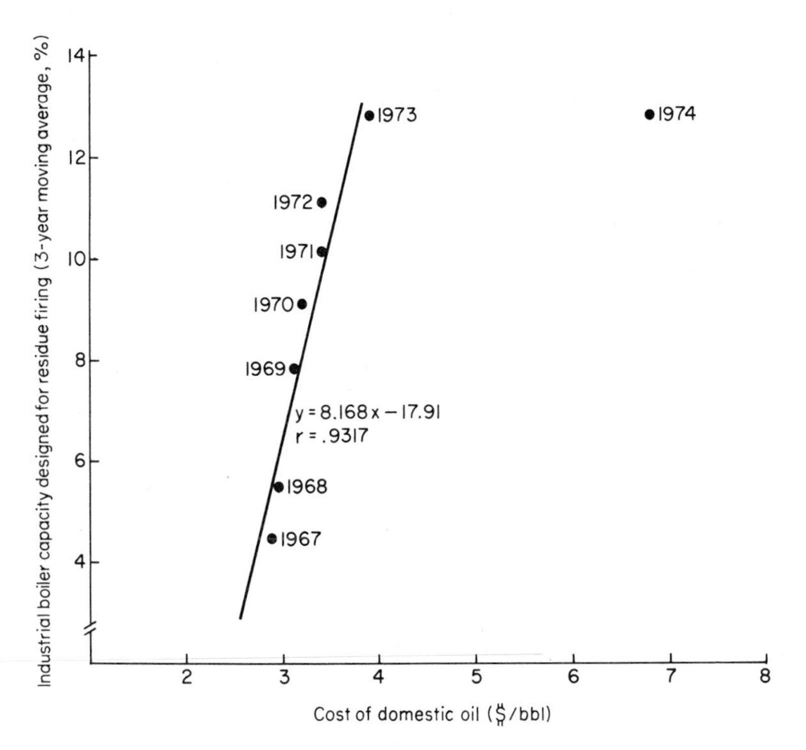

FIGURE 2. The influence of oil prices on residue fired industrial boiler capacity.

have combined with the capital crunch to obviate the possibility of reviving that trend in a new position. Consequently, we can observe that oil price played a significant role at the start of the buildup, but its present role is unknown beyond intuitive observations.

Absolute energy availability appears to be more of a question for the future, one that subsequent papers will deal with. Certainly domestic oil and gas production has peaked. Oil and gas imports from Canada are declining, while imports from the Persian Gulf nations are increasing. This phenomenon, however, is very recent and not yet amenable to statistical analysis. Absolute energy shortfalls did shut down many businesses in 1976, however. Thus, Peabody Gordon-Piatt placed a quarter page advertisement in the Wall Street Journal in March, 1977, with the following headline: "Stay in business during fuel curtailments! Burn plant wastes instead." Obviously, they see absolute fuel availability as a motivating force. Anticipation of future shortfalls could spur a new round of increasing installations designed to be fueled with nonfossil organic materials.

2. Environmental Protection Reasons

The recent nine-year trend in the increased utilization of nonfossil organics, supported by a growing number of boiler installations designed for residue firing, is as associated with environmental considerations as it is with energy prices. During the early and mid-1960s, it was not unusual to see an oil-fired boiler installed next to a wood-residue-destroying incinerator. Combustion of residues without concern for energy recovery, was the rule when fuel was cheap and incineration easy. With the advent of the Clean Air Act and the need to clean up incinerators, the combination of environmental credits plus modestly increasing fuel oil prices made energy recovery more attractive. Subsequent amendments to that act, plus the Clean Water Act, reinforced the trend.

The correspondence between environmental laws and increases in the residue-fired industrial boiler share of the market is more than coincidental. Figure 3 depicts the relationship between the trend and the laws. Significantly, the two three-year plateaus are closely tied to the Clean Air Act Amendments and the Clean Water Act. The initial buildup appears to have resulted from the Clean Air Act itself. When one considers that of the fuel consumed by industry, about 1.70 quads is residue generated and burned by the same company, 0.102 quads is residue sold to another company for its utilization as fuel, and 0.0005 quads is material harvested and consumed specifically for fuel, then the observation is reinforced.

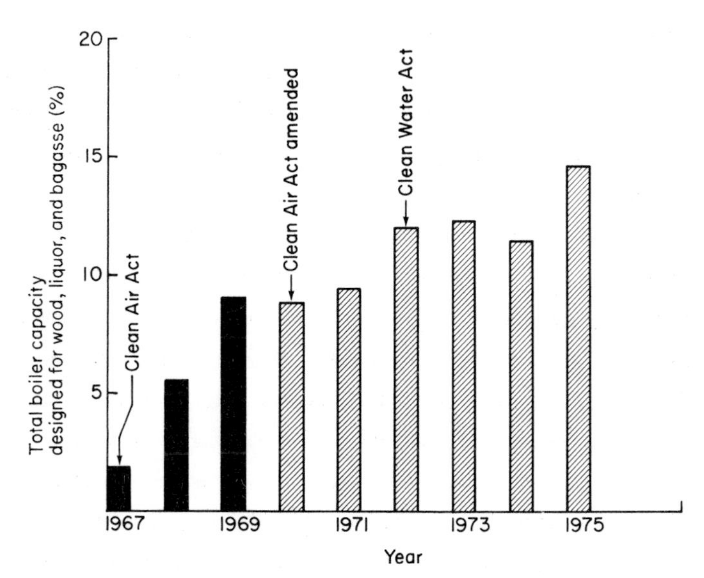

FIGURE 3. Trends in the installed capacity of wood, spent liquor, and bagasse fired boilers for industry by legislative period.

Spokesmen for both Riley Stoker and Combustion Engineering confirm this synergism between environmental regulations and energy credits as the principal force. Mr. Al Downham of Riley Stoker confirmed this in remarks quoted by Energy User News: "It's like killing two birds with one stone. We have to dispose of it (waste) some way, so we might as well use it. We are using what was an otherwise undesirable material. Industries that generate combustible materials in quantity are most interested. They are in a critical situation. Fuel costs have gone up so high, the value of refuse is now up also [49].

V. CONCLUSION

Nonfossil organic fuels were mankind's first energy resource and throughout history have played a prominent role. They fueled the U.S. economy during its initial period of industrialization, losing out to coal in 1880. Now both nonfossil organics and coal are growing. At this point renewable organic resources are contributing 1.8 quads of energy per year. Their annual growth rate is over 5%.

Two synergistic forces have combined to create the resurgence of nonfossil organics: energy prices and environmental protection regulations. This is true not only for wood-based materials but also for urban waste. More forces are now emerging to spur the growth on. Absolute energy availability (scarcity), Federal energy conservation regulations of a formal and informal nature, and expectations of future energy-related events are among these new factors. Future analysis may determine their role in increasing energy production from renewable organic resources.

ACKNOWLEDGMENTS

The author wishes to acknowledge the invaluable assistance of Mr. William Axtman, American Boiler Manufacturers Assn., for providing historical trends in boiler installations. The author also wishes to acknowledge the assistance of three key reviewers: Dr. Fred Shafizadeh, University of Montana; Dr. K. V. Sarkanen, University of Washington; and Dr. Bernard Blaustein, Pittsburgh Energy Research Center, ERDA.

REFERENCES

1. Fred Shafizadeh and William F. DeGroot, "Combustion Characteristics of Cellulosic Fuels," Thermal Uses and Properties of Carbohydrates and Lignins. Academic Press, Inc., 1976.
2. Herman Kahn, et al, The Next 200 Years, William Morrow & Co., Inc., 1976.
3. H. G. Cordero and L. H. Tarring, Babylon to Birmingham. Quinn Press Ltd., London, England.
4. Nathan Rosenberg, Technology and American Economic Growth. Harper and Row, 1972.
5. Charles J. Jordan and Jessie S. Cole, "The Shape of Things to Come." Yankee, Jan. 1974.
6. Sam H. Schurr and Bruce C. Netschert, Energy in the American Economy, 1850-1975. John Hopkins Press, 1960.
7. E. C. Wiley, "Rating and Care of Domestic Sawdust Burners," Bulletin Series, No. 15, Engineering Station, Oregon State College.
8. Chalmer G. Kirkbride, "ERDA Projects on Biomass Conversion," Capturing the Sun Through Biomass Conversion, Proceedings, March 10-12, 1976.

9. R. G. Yeck, "Panel Discussion Remarks," Proceedings-International Biomass Energy Conference, May 13-15, 1973.
10. Russell W. Peterson, "The Ecology of Bioconversion," Capturing the Sun Through Bioconversion, Proceedings, March 10-12, 1976.
11. John Heslop-Harrison, "Reforming the Cellulose Economy," New Scientist, Jan. 30, 1975.
12. J. M. Duke and M. J. Fudali, "Report on the Pulp and Paper Industry's Energy Savings and Changing Fuel Mix." American Paper Institute, Sept., 1976.
13. Scott Minerbrook, "Chemical Recovery Boiler to Cut Oil Use," Energy User News, Dec. 27, 1976.
14. "New Hampshire Paper Mill Utilizes Nearby Town's Refuse. Catalyst, Vol. V., No. 3, 1976.
15. "Firm in Alabama Begins Building Huge Pulp Mill." Wall Street Journal, Nov. 23, 1976.
16. U. S. Forest Service, "The Feasibility of Utilizing Forest Residues for Energy and Chemicals." National Science Foundation, March, 1976.
17. "Wood Chips Make Industrial Gas." Energy User News, Jan. 10, 1977.
18. L. H. Banning and W. E. Anable, "Preliminary Electric Smelting Research on Philippine Nickeliferous Ores," U. S. Bureau of Mines R.I. 5219, May, 1955.
19. "The Hanna Nickel Operation." The Hanna Mining Company, Hanna Nickel Smelting Co., Riddle, Oregon, June 1, 1970.
20. White Pine Copper Smelter Superintendent, personal conversation, December, 1976.
21. Owen D. Brown, "Energy Generation from Wood Waste," International District Heating Assn., French Lick, Indiana, June 20, 1973.
22. "Textile Mill Uses Woodwastes for Power." Catalyst, Vol. V, No. 3, 1976.

23. L. Fletcher Prouty, AMTRAK Corp., correspondence Oct. 12, 1976.
24. Ed Cliff, Timber, the Renewable Resource. National Commission on Materials Policy, 1973.
25. "OSU Studies How Fast Wood Burns." Albany Democrat-Herald, Albany, Ore., Dec. 26, 1976.
26. U. S. Bureau of Census, Battelle Columbus Laboratory, Evaluation of National Boiler Inventory, Oct. 1975.
27. Mary Ellen Perry, "Warming Up To Those Old Wood Stoves." The Washington Star, Feb. 5, 1977.
28. Arthur Seeds, Charcoal Briquette Institute, interview Dec. 28, 1976.
29. William Arlington, "Bagasse As a Renewable Energy Source," Fuels and Energy From Renewable Resources. Academic Press, 1977.
30. W. F. Lalor, J. K. Jones and G. A. Slater, "Performance Test of Heat-Recovering Gin-Waste Incinerator," Agro - Industrial Report, Cotton Inc., 1976.
31. Sharon and James Whitehurst, "Our Four Cow Biogas Plant," Carol H. Stoner (ed), Producing Your Own Power. Rodale Press, 1974.
32. A. Roger Grout, "Methane Gas Generation from Manure." Cooperative Extension Service, Pennsylvania State University, 1974.
33. Frederick T. Varani and John J. Burford, Jr., "The Conversion of Feedlot Wastes into Pipeline Gas, Fuels From Waste, Academic Press, 1977.
34. David A. Tillman, "At Monfort of Colorado, Inc.--Vertical Expansion is Key to Success." Area Development, Jan. 1976.
35. L. L. Anderson, "Energy Potential From Organic Wastes: A Review of the Quantities and Sources." U. S. Bureau of Mines IC 8549, 1972.
36. "Companies Mine Energy From Their Trash." Business Week, August 2, 1976.

37. Richard A. Young and Ian O. Lisk, "Kodak Disposes of Waste in System That Helps Pay Its Way." Pollution Engineering, Sept. 1976.

38. "How Trash Is Being Turned Into Useful Heat." Environmental Science and Technology, Sept. 1976.

39. E. R. Moats, "Goodyear Tire Fired Boiler." Journal of the Washington Academy of Sciences, Vol. 66, No. 1, March, 1976.

40. "Companies Mine Energy From Their Trash." Business Week, August 2, 1976.

41. Carol Frey, "Deere Shrugs Off Weaker Payback for Solar Heater," Energy User News, March 7, 1977.

42. Kjell I. Erlandsson, "Using Solid Waste as a Fuel," Plant Engineering, Dec. 11, 1975.

43. Robert G. Schweiger, "Power From Waste," Power, February 1975.

44. William Axtman, American Boilers Manufacturers Association, personal interview, November 2, 1976.

45. Robert H. Collins, III, "Gas Recovery: National Potential." Waste Management Technology and Resource and Energy Recovery, U.S. Environmental Protection Agency, 1976.

46. John Pacey, "Methane Gas in Landfills: Liability or Asset?" Waste Management Technology and Resource and Energy Recovery, U.S. Environmental Protection Agency, 1976.

47. "Sheldon-Arleta Gas Recovery Project," City of Los Angeles Department of Public Works, Bureau of Sanitation, Mar. 1976.

48. National Commission on Water Quality on Assessment of Technologies and Costs, Vol. 2, Metcalf & Eddy, Inc.

49. Alan Dell, "Waste-Fired Boiler Firms Gird for Booming Market," Energy User News, March 14, 1977.

ANTICIPATED COMPETITION FOR AVAILABLE WOOD FUELS IN THE UNITED STATES

John B. Grantham

Pacific Northwest Forest and Range Experiment Station
Seattle, Washington

I. INTRODUCTION

This paper presents an overview of the competition that may develop for wood as an energy source vs. its use as an industrial raw material. Competition is anticipated in filling future fuel and fiber needs from unused wood residues or from unmerchantable timber and the ouput of energy plantations. As a basis for this overview, current and potential growth of timber in the United States and the current and projected demands on the timber resource in 2000 are reviewed.

Special attention is given to a plausible distribution of available wood supplies among major segments of the forest products industries, based on the anticipated needs of those industry segments by the year 2000. This special attention is justified because the forest industries have several advantages over other industries in using wood or bark for either raw material or energy, including: (a) proximity to wood supplies; (b) experience in handling bulky, wet, and sometimes dirty material; (c) ability to convert wood to a mix of products or electrical and mechanical energy at relatively high efficiencies; and (d) a critical need to become independent of interruptible natural gas.

Wood is renewable and versatile and may become the world's most important basic raw material within the next century [1]. Meanwhile, there are surpluses that could contribute to some of the Nation's short-term energy needs.

Tillman [2] has identified three potential constraints to greater use of organic residues for energy in the United States. These are: (a) the present rate of residue use; (b) competition for residues; and (c) capital costs associated with burning residues. This paper addresses the first two.

II. CURRENT AND PROJECTED WOOD PRODUCTION IN THE UNITED STATES

Forests occupy about one third of the country's land area. Two thirds of this area or 500 million acres (202×10^6 ha) is classed as commercial forest land. This base is changing with additional withdrawals for a variety of uses; and by 2020, it is estimated that the commercial forest land area will be reduced to between 455 million and 475 million acres [3].

Net timber* growth in the United States is estimated to be about one-half cord per acre per year. Based on Sweden's experience with intensively managed forests, timber growth could be

---

* Timber is defined as the stemwood of merchantable trees measured from 1 ft above ground to a 4-in. top diameter. All trees over 5 in. in diameter at breast height are included. Stemwood makes up 80-85% of above ground biomass in softwood trees. Stemwood and large branchwood make up 80-90% of hardwood tree biomass.

doubled to 0.95 cord per acre ($5.3m^3$/ha) per year by 2020 [3]. This projection is based on estimated yields of fully stocked timber stands. These estimates are adjusted for withdrawals of forest lands; for plus factors such as weeding and thinning, fertilization and drainage, genetic improvement, and conversion of forest types; and for negative factors such as unrecovered mortality and holes in the forest stand.

The analysis of potential timber growth by Spurr and Vaux [3] also considers several constraints:

> Although management intensity on the 17 percent of land owned by forest industries is expected to achieve full potential growth, no significant increase is expected on the 58 percent of commercial forest land held by nonindustry private owners. Forestry investments by this ownership class are commonly discouraged by inefficient size of ownership, meager access to capital, uninsurable risks, and adverse local tax systems. Additionally, an estimated 15 percent of these owners hold their lands for purposes generally incompatible with timber growing.
>
> On the 25 percent of commercial forest land in public ownership, some timber-growing potential will be forgone to achieve other resource use objectives (recreation, fisheries, etc.).

In conclusion, the authors estimate that the biological potential of fully stocked stands is an estimated 432 million cords per year. The economic potential, however, is believed to be 15% less or 367 million cords, and the economic potential under existing institutional constraints is believed to be about 238 million cords per year.

Detailed projections of timber supply and demand prepared by the U.S. Forest Service [4] anticipate only a modest increase in net growth of from 232 million cords per year in 1970 to 245 million cords per year in 2000. This projection, based on the 1970 level of forest management, indicates that resource management must be intensified greatly if we are to achieve potential growth on even the better growing sites.

Another concern of forest resource managers is that 58% of the commercial forest land (75% of the hardwood type stands) is owned by several million farmers or miscellaneous private owners. Only about one third of these owners have some interest in forestry, and perhaps only 5% manage their forest lands intensively [3]. Thus, it is vitally important to enlist a greater interest among these land owners in the Nation's projected need for wood fiber and fuel.

III. CURRENT AND PROJECTED USE OF DOMESTIC WOOD

Estimated use of domestic roundwood in 1970 was 160 million cords, not a great overall change from the average consumption of past years, as illustrated in Table I. During the past 120 years, however, the use of roundwood for fuel has declined from major to minor importance [5], but its use for pulp has increased steadily.

Each of the six use classes for domestic roundwood shown in Table I requires some special considerations to estimate current and future demand. For this reason, the estimates of roundwood and residue demand for each use class are considered separately. With these considerations, the current and projected use of deomstic wood is summarized in Tables VI-A and VI-B.

A. Fuelwood

Worldwide fuelwood still accounts for almost half of roundwood use [1] or more than 500 million cords--more than three times the total use of roundwood in the United States. Fuelwood

TABLE I. *Domestic Roundwood Use for Products and Exports [4,5,7] (In Millions of Cords)[a]*

Year	Fuel[b]	Lumber[c]	Pulpwood[d]	Plywood[e]	Misc. products	Exports	Total[f]
1850	102						
1875	137						
1900	100						
1920	83	70.0	4.7				155
1930	75	58.8	5.2				139
1940	70	62.4	12.1				145
1950	60	76.0	19.5		9.0	0.1	165
1960	35	65.8	33.5	7.9	6.6	0.55	149
1965	15	73.6	40.3	12.8	5.8	2.4	128
1970	7	77.8	50.2	14.1	5.0	5.5	160

a. 1 cord = 80 cu ft of wolid wood or 2200 lbs/cord for softwood; 2600 lbs/cord for hardwoods.
b. Does not include residues from other industries used as industrial fuel, est. at 9MM cords.
c. 500 bf = 1 cord of roundwood d. Pulpwood does not include residues used.
e. 1128 sq ft (3/8" basis) = 1 cord of roundwood.
f. Fifteen to twenty million cords of logging residues should be added to the annual drain on available timber.

NOTE: Of 72 million cords of residue developed in lumber and plywood manufacture, 28 million cords were used for pulp production, 20 million cords for fuel, and 24 million cords were unused in 1970.

use has grown by 10% in the past decade, and industrial wood use has grown by 30%.

Roundwood use for fuel in the United States decreased by 1970 to less than 5% of its one time use (Table I). Tillman [2] believes that 1970 was the low point of wood fuel use in this country and that its use will increase as people rely on it for a supplementary fuel. This point is made in his paper presented in this volume.

Use of wood and bark residue for fuel, which amounted to an estimated 9 million cords in 1970, is expected to increase substantially but will be limited almost entirely to the forest products industries to fill their energy needs. Pressed dry logs or pelletized fuel from residue will find some favor but these represent relatively inefficient conversion to heat as compared with burning wood residue in industrial furnaces, especially at their point of origin.

Increased domestic use of roundwood for fuel is forecast to supplement other heat sources. This increase, which could amount to 20-25 million cords per year by 2000, is expected to come largely from underutilized hardwood stands.

B. Lumber Manufacture

1. The Current Situation

Domestic lumber production required an estimated 77.8 million cords of roundwood in 1970 to produce 34.7 billion board feet of lumber. Domestic use of lumber in that year was estimated at 39.5 billion board feet with lumber imports exceeding exports by 4.8 billion board feet.

If the 61 million cords of softwood logs and 17 million cords of hardwood logs and accompanying bark (98 million tons) is reduced to lumber; the equivalent of 65 million tons, oven-dry (60% of the roundwood input), and including an estimated 10 million tons of bark emerges as residue (Table II). An estimated

TABLE II. Current and Projected Disposition of the Residues Generated in Lumber Manufacture -1970 and 2000 (Millions of tons, oven-dry)

	1970	2000	Change
Roundwood input[a]	98	103	+ 5
Primary product	33	42	+ 9
Residue generated	65	61	- 4
Byproducts to others	21	32	+11
Fuel (in plant)	6	22	+16
Fuel (to others)	12	7	- 5
Unused residue	26	0	-26
Total	65	61	- 4

a. *Added requirements: 5 million tons of hardwood logs.*

18 million cords (21 million tons) of the wood residue is used for products such as pulp and particleboard, leaving 44 million bone dry tons* available for fuel.

In 1970, it was estimated that about 18 million tons, roughly half of this available residue, was used for fuel, largely in forest industry plants and usually at the plants where it originated. The balance of 26 million tons was available for use outside the industry at a nominal price of $1.00 to $2.50 per oven-dry ton.

* Based on 1 cord = 80 cu ft at 30 lbs/cu ft or 1.2 short ton, oven-dry (1.09 metric ton) per cord. (Softwood at 1.1 oven-dry ton/cord; hardwoods at 1.3 oven-dry tons/cord.)

2. Future Trends

It is estimated that roundwood use for lumber will increase 5-25% in the 30-year period 1970 to 2000, depending on numerous factors, such as the relative rise in lumber prices with respect to other building materials. In anticipation of strong competition for softwood timber, it is believed that the 5% figure is more nearly correct and that nearly all of the increase will be hardwood lumber. This will require an estimated 61 million cords of softwoods and 21 million cords of hardwoods or 103 million tons of log input, including bark. This anticipates, too, that the United States will continue to rely on imports for a substantial amount of its softwood lumber needs.

The increasing value of timber and the wider adoption of improved technology should raise lumber recovery to 45% of the material input for softwoods and 30% for hardwoods, leaving 61 million tons available as residue. Current and projected disposition of these residues is shown in Table II.

In its drive for energy self-sufficiency to insure against operational shutdowns, the lumber industry is expected to generate all of its heat requirements and a portion of its electric power requirements from manufacturing residue. Energy requirements are projected at 3.7 million Btu's per ton of logs processed, based on estimates by Boyd, et al. [6] of 1400 pounds of process steam and 62.8 hp-hours per oven-dry ton of current log input. For the year 2000, the requirements were modified to project higher lumber recovery per ton.*

The 103 million tons, oven-dry of log input will require an estimated .383 Quads (quadrillion British thermal units) of

---

* Anticipated output per ton, oven-dry, of softwood log input:

Planed lumber	0.45 ton
Pulp chips	0.20 ton
Particleboard furnish	0.14 ton
Fuel	0.21 ton
Total	1.00 ton

energy for conversion to lumber and byproducts. Twenty-two million tons of residue for fuel could provide .383 Quads at 17.5 million Btu/ton, oven-dry. This leaves a balance of 11 million tons available to other industries.

Although 23 million tons of fuel (.40 Quads) were surplus to the lumber industry in 1970, it is anticipated that by 2000 all residue fuel generated will be needed to supply energy requirements of the forest products industries regardless of the quantity of lumber produced domestically.

C. Plywood Manufacturers

1. Current Situation

In 1970, an estimated 14.1 million cords of roundwood were used to produce 14.1 billion sq ft (3/8 in. thickness) of softwood and 1.8 billion square feet of hardwood plywood [6].

Using the conversion factors reported by Boyd et al [6], the roundwood input of 12.5 million cords of softwood (15.1 million tons, oven-dry) and 1.57 million cords of hardwood (2.25 million tons, oven-dry) will produce 7.4 million tons of byproducts and 2.33 million tons of oven-dry fuel* for energy production. Currently, more than 60% of the bark and fine residue is used as fuel in the plywood industry, leaving some 920,000 tons available (Table III).

2. Future Trends

By the year 2000, it is estimated that roundwood input for plywood manufacture and structural flakeboard will increase from 14 million to 25 million cords (20 million cords for plywood; 5 million cords for flakeboard. If we assume that the portions of primary products, byproducts, and fuel produced in converting

* 1 cord = 1.1 short ton, oven-dry for softwoods, plus 10% bark or 1.21 ton. 1 cord = 1.3 short ton, oven-dry for hardwoods, plus 10% bark or 1.43 ton.

TABLE III. Current and Projected Disposition of the Residues Generated in Plywood Manufactur - 1970 and 2000. (Millions of tons, oven-dry)

	1970	2000	Change
Roundwood input[a]	17.4	24.2	+ 6.8
Primary product	7.7	9.9	+ 2.2
Residue generated	9.7	14.3	+ 4.6
Byproducts to others	7.4	8.2	+ 0.8
Fuel (in plant)	0.2	6.1	+ 5.9
Fuel (to others)	1.2	0.0	- 1.2
Unused	0.9	0.0	- 0.9
Total	9.7	14.3	+ 4.6

a. Added requirements - 6.8 million tons - 5.1 softwood, 1.7 hardwood.

roundwood to plywood will remain much the same as at present, the following quantities are anticipated: 15 million cords of softwoods (75%) and 5 million cords of hardwoods should produce about 6.1 million tons of fuel. If current industry practice continues, the plywood industry should also provide 8.2 million tons of pulp chips by the year 2000.

The plywood industry, like the lumber industry, has modest energy requirements. However, as the industry moves toward energy self-sufficiency, it will be necessary to reduce slightly its commitment of raw materials to the pulp and particleboard industries. If we assume that the majority of plywood plants will provide heat for drying but little of their electric power demand, a factor of 3.5 million Btu's per ton of roundwood input is anticipated (vs. 3.7 million Btu's per ton for conversion to lumber). This energy requirement is equivalent to 0.2 ton of residue. At an industry average of .15 ton of oven-dry fuel per ton of log input, there is a shortage of .05 ton industry-wide, roughly equivalent to the amount of particleboard furnish supplied by softwood sheathing plants. Thus, to achieve energy self-sufficiency, additional material must be diverted from chips or particleboard furnish to fuel, as indicated in Table III. Overall, the industry is expected to require .09 Quad of energy with no anticipated surplus of fuel.

D. Pulp Manufacture

1. *Current Situation*

In 1972, the U.S. pulp industry used 48 million cords (55 million tons, oven-dry) of domestic roundwood and the equivalent of 25 million cords (29 million tons) of chips from lumber, plywood, and miscellaneous industry byproducts to produce 42 million tons, oven-dry, of wood pulp and 55.4 million tons of paper and paperboard products [7]. In doing so, the industry obtained an estimated 35% of its energy needs from black liquor,* bark, and a small quantity of purchased hogged fuel (Table IV). Sixty-five percent of the estimated energy required was purchased or the equivalent of 264 million barrels of oil.

2. *Future Trends (Fiber Needs)*

Future world and North American demand for paper and paperboard has been estimated by Hagemeyer [8]. When all demand factors, as well as constraints are considered, Hagemeyer's prediction of North American pulp consumption in the year 2000 varies from 100 to 120 million tons, oven-dry, compared with a 1975

---

* Black liquor is the spent cooking liquor from alkaline pulping containing inorganic chemicals and dissolved organic components of pulpwood. To recycle the inorganic chemicals, the liquor is concentrated and burned, thereby permitting the recovery of heat from the dissolved organics.

TABLE IV. Current and Projected Wood and Bark Requirements of the Pulp and Paper Industry - 1970 and 2000. (Millions of tons, oven- dry.)

	1970	2000	Change
Inputs for fiber			
Roundwood	57.5	69.0	+11.5
Chips from lumber, plywood manufacture	23.0	32.0	+ 9.0
Chips from other materials, including whole tree chips	-	44.0	+44.0
Total	80.5	145.0	+64.5
Inputs for fuel			
Black liquor and bark[a]	60.0	104.0	+ 44.0
Purchased hogged fuel	0.5	112.5	+112.0
Total	60.5	216.5	+156.0
Plus oil or equivalent fuel purchased (in millions barrels)	264[b]	30[c]	-234

a. It is estimated that roundwood acquisitions will include about 6 million tons of bark suitable for fuel.

b. To provide the balance of industry's energy needs.

c. Assume 10% minimum of oil or other fossil fuel for startup, standby, etc.

consumption of 64.5 tons. The American Paper Institute expects overall wood pulp needs of the U.S. pulp and paper industry to double by 2000, or to reach a level of about 80 million tons, oven-dry.

Auchter [7], in discussing raw material supplies for the industry, predicts that the fiber sources for wood pulp will change rather dramatically between 1974 and 2000 (Table V). In essence, he projects a substantial increase in the use of logging residue and in whole tree chips from forest stands. These projections are based, in part, on anticipation that manufacturing residues will be increasingly used for energy within the lumber and plywood industries.

TABLE V. Current and Projected Fiber Sources for Wood Pulp [7] *(Millions of tons, oven-dry)*[a]

	1974	*2000*
Roundwood	*52*	*69*
Manufacturing residue	*27*	*32*
Logging residue	*3*	*16*
Urban residue	< *0.1*	*5*
Whole tree chips	*1*	*23*
Pulpwood imports[b]	*1*	-
Totals	*84*	*145*

a. *Mix of softwood and hardwood at 1.15 tons/cord.*

b. *Anticipates importation of pulp rather than pulpwood.*

3. *Future Trends (Energy Needs)*

The paper and paperboard industry has an enormous energy requirement, varying greatly with the type of product being produced. Slinn [9] reports an industry average of 32.3 million Btu's per short ton, air-dry, for the first 11 months of 1975 but a decrease to 31.2 million Btu's per ton for the last 5 months of the period. If we assume that the industry's drive for energy conservation will yield additional savings, a requirement of 30 million Btu per short ton is projected. This figure is only slightly less than current experience in Scandinavia, where the high energy costs of the past have emphasized a need for energy-efficient operation.

Slinn reports that in a typical integrated, bleached kraft pulp and paperboard mill, the energy use is:

Pulping	41%
Paper forming	41%
Bleaching	17%
Debarking and chipping	1%

His report also summarizes several potential changes in technology that can affect energy efficiency but cautions that two major constraints can delay adoption of new energy-efficient technology. These are:

a. high capital costs and potentially inadequate return often limit the ability of individual companies to adopt new technologies;

b. measures taken to reduce pollution increase energy requirements and reduce the availability of capital to adopt more energy-efficient processes.

Auchter's forecast [7] of pulpwood sources for 2000 indicates a total of 126 million cords or about 145 million tons, oven-dry (Table V). This would provide 72.5 million tons, oven-dry (80 million air-dry tons) of wood pulp at an average yield of 50%. If we assume roughly the same ratio of paper and paperboard production to domestic wood pulp production (1.32 to 1.0) as prevailed in 1972-1976, this should mean a U.S. industry output of 96 million tons and an energy requirement of 2.88 Quads. If internal energy sources provide 45 to 50% of requirements, there will still be need for about 1.35 Quads of additional energy or a 45% increase over current needs.

The above is equivalent to 300 million barrels of oil or 125 million tons of wood and bark. It is assumed that the industry will use at least 10% or some 30 million barrels equivalent of fossil fuel for startup, standby, etc. leaving a need for about 112 million tons of wood or bark fuel, if available.

Because of the close association between lumber, plywood, and wood pulp manufacturing plants, often under one ownership, the pulp industry may be expected to satisfy much of its energy needs with wood and bark residues available from lumber manufacturing plants or logging operations. It is unlikely, however, that much of the increased need can be supplied by manufacturing residue, especially if the pulp industry increases its pulp chip purchases from lumber and plywood manufacturers as predicted in Tables VI-A and VI-B. In brief, the paper and paperboard industry can use all surplus wood and bark residue available in 2000 without filling its need. Thus, the industry can be expected to consider several of the sources for fuel as well as fiber (see Table V).

E. Miscellaneous Products

1. Current Situation

Miscellaneous products include conventional particleboard; such raw products as poles, posts, piling, and mine timbers; and such manufactured products as shingles, shakes, cooperage, and charcoal. In total, these products use a declining amount of roundwood but an increasing quantity of manufacturing residue. In 1970, each of the three groups of products mentioned above required about 2.5 million cords of raw material. Nearly all the material used in particleboard production was manufacturing residue of the lumber and plywood industries. The other miscellaneous products used roundwood almost exclusively.

2. Future Trends

The quantity of roundwood used for miscellaneous products is expected to stabilize at about 5 million cords per year. The quantity of wood used for conventional particleboard manufacture is expected to grow by more than 50%, whereas a new structural particleboard should consume an estimated 5 million cords of roundwood.

Overall, it is anticipated that miscellaneous products may require 10 million cords of roundwood for structural particleboard and all other products, plus at least 5 million cords of wood residue for conventional particleboard by 2000. Energy requirements for particleboard production are estimated to be 0.042 Quads, which is equivalent to another 2.5 million tons of residues.

F. Export Wood

A considerable quantity of logs, milled cants, and pulp chips are exported from Pacific Coast ports, chiefly to Japan. In 1970, log exports amounted to 5.5 million cords (6.6 oven-dry tons) [4]. Chip exports in the same year amounted to 2.1 oven-dry tons [10]. Although it may be reasoned that this quantity of raw material can be available to domestic producers when needed, export logs and export chips generally have commanded higher prices than those offered by domestic manufacturers. This Pacific Coast export market has provided higher returns to support log harvesting and other forestry operations. These exports are more than offset by lumber and pulp imports from Canada, which are equivalent to some 17 million cords of roundwood annually.

Overall, it is anticipated that softwood log and chip exports will remain at about the current level despite Japan's growing need for wood and its preference for North American softwoods. Although there is a possibility of shifting some pulp chip exports from softwood to hardwood, there is little chance of exporting hardwood logs or cants.

TABLE VI-A. *Roundwood Requirements and Production and Use of Wood and Bark Residues in the Forest Products Industries - 1970*

Use	Roundwood required		Residue generated	Residue used		Residue unused
	millions cords	millions tons, oven-dry		for products	for fuel	
			millions of short tons, oven-dry			
Lumber	77.8	98.0[a]	65.0	(21.0)[b]	6.0 (12.0)	26.0
Plywood	14.1	17.4	9.7	(7.4)	0.2 (1.2)	0.9
Pulp and paper	50.2	63.5	44.0[c]	23.0[d]	44.0 0.5[e]	None
Misc. products	5.0	5.7	2.5	3.4[f]	None	2.5
Fuelwood	7.0	8.4	None		12.7	
Export logs	5.5	6.0	None			
Export chips				2.0		
Totals	159.6	199.0	121.2	28.4	63.4	29.4

a. Wood × 1.15 short ton, oven-dry/cord + 9 million tons of bark (10%).

b. Numbers in parentheses are quantities used elsewhere by others., e.g. lumber and plywood residue used for pulp and paper, miscellaneous products and export chips.

c. Spent black cooking liquor solids (organic and inorganic) plus 5.8 million tons bark.

d. Purchased chips from lumber and plywood residues.

e. Purchased hogged fuel.

f. Residue purchased for particleboard.

TABLE VI-B. Projected Roundwood Requirements and Production and Use of Wood and Bark Residues in the Forest Products Industries - 2000

Use	Roundwood required		Residue generated	Residue used	
	millions cords	millions tons oven-dry		for products	for fuel
			(millions of short tons, oven-dry		
Lumber	82	103.0	61.0	(32.3)[a]	22.0 (6.7)
Plywood	20	24.2	14.3	(8.2)	6.1
Pulp	60	75.0[b]	104.0[c]	32.0[d] 44.0	104.0 5.5[e]
Misc. products	10	12.6	2.5	5.5[f]	2.5 1.2[e]
Fuelwood	20	25.0	None		
Export logs	5	5.5	None		
Export chips				3.0	
Totals	197	245.3	181.8	84.5	141.3
Additional residue from urban waste, whole tree chips, etc.			44.0		
			225.8		

a. Numbers in parentheses are quantities available to others, e.g., lumber and plywood residue for pulp and paper, miscellaneous products, export chips, and hogged fuel.

b. Includes an estimated 6 million tons of bark.

c. The combined quantity of roundwood, manufacturing residue, and whole tree chips (151 million tons) will produce an estimated 98 tons, oven-dry, of black liquor solids and 6 million tons of bark for energy.

d. Purchased residue; includes manufacturing residue 32 million tons; whole tree chips, etc. 44 million tons.

e. Hogged fuel from lumber manufacturing residue.

f. Particleboard furnish from urban or forest residue.

IV. SOURCES AND ESTIMATED AVAILABILITY OF WOOD FUELS

A. Manufacturing Residues

1. *Lumber*

The current availability of surplus residue from the lumber industry is estimated to be 23 million tons, oven-dry. This surplus residue is declining, however, as the lumber industry uses an increasing amount for fuel to satisfy its own needs (Table II). Whatever residue, primarily bark, is available, or surplus to the industry, will be in increasing demand by the pulp and particleboard industries, both for fiber and for fuel, although there will be some local movement of fuel to others, as hogged fuel, compressed logs, pellets, etc. In turn, the industry will need some supplementary fossil fuel for startup and to meet surges in demand.

2. *Plywood*

As discussed earlier, the plywood industry has been an important supplier of wood residue to the fiber industries and has had a surplus of bark or other materials for fuel. Studies indicate, however, that in order for the plywood industry to be energy self-sufficient, it will need to reduce its commitment of residue raw materials to the pulp and particleboard industries by about 20%.

3. *Pulp and Paper*

The pulp industry has two manufacturing residues that are used to generate energy within the industry. The first is the bark from pulpwood that is debarked at the pulpmill. It is burned along with other fuels in the power boiler and in 1972 provided 5% of the industry's energy needs [9].

The second, and more important residue, spent cooking liquor, which contains about half the weight of wood, can provide nearly all energy requirements in an unbleached kraft mill.

Overall, cooking liquors supplied 32% of the industry's total energy needs in 1972 [9].

4. Miscellaneous Products

Producers of poles, piling, shakes, shingles, and other miscellaneous products do have a surplus of residue, particularly on the Pacific Coast, but the amounts are small compared with overall fiber and fuel requirements of the forest products industries. Because producers of particleboard are included in this class of products, overall the group requires more residue than it produces. As a class, it ranks as an important user of residue, primarily from the lumber and plywood industries.

5. Logging Residues

Logging residue has been considered as a source of energy by Grantham and Ellis [11] and is estimated to amount to at least 35 million tons, oven-dry, per year. Because of the wide variability in the tonnages of residue left in logging, it is difficult to estimate the total quantity accurately. Furthermore, the residue is scattered over wide areas, often long distances from manufacturing centers. Because of remoteness, not more than three quarters of the material accumulating each year is likely to be recovered.

Because the bulk of logging residue is suitable for fiber products, as well as for fuel, more than half of any recovered residue may be unavailable for fuel. The added return from products other than fuel, however, can help offset the high cost of delivered residue and can make recovery economically feasible.

Adams [12] has recently described the role of the whole log chip mills in converting the larger pieces (cull logs) of logging residue to pulp chips. Although operation of these mills is dependent on the condition of the world pulp market, the mills will continue to play a key role in the use of residue on the Pacific Coast. Each whole logs chip mill represents a

$3 million investment and can process 60 tons (wet) of logs per hour. In converting large quantities of logging residue to chips, these mills develop fuel from the bark and fines amounting to 12-15% of the input (8 tons/hr). Adams adds that, because of the cyclic nature of the chip market, it is important that the chip mills be owned by integrated forest products companies.

6. *Underutilized Stands*

A major problem in providing for the Nation's projected industrial raw material needs is the relative demand for softwoods and hardwoods. In general, softwood species are preferred for the major wood uses, including lumber, plywood, paper, and paperboard. This results in a demand for softwood timber that is more than double that for hardwoods.

In 1970, softwood removals were 120 million cords (90% of estimated softwood growth); hardwood removals were 55 million cords (56% of estimated hardwood growth). In view of projected increases in demand for wood as an industrial raw material, there is need to improve the balance between softwood and hardwood use. This is further emphasized by the fact that more than half the commercial forest now supports hardwood growth. On non-industry forest lands (or those in smaller private ownership) where better management is critically needed, over 70% of the lands support hardwood timber types.

Hardwood timber types are well suited to supplying energy. Although their heat value per pound of wood, oven-dry, is less than resinous softwoods, their density (weight per cubic foot) is generally high enough to provide a higher heat value per cord.

In view of the disparity in the supply and demand of hardwood timbers, both the United States and Japan have devoted considerable effort to improving the characteristics of hardwood pulps, or pulp blends, so that an increasing portion of the growing pulp demand can be satisfied by hardwoods. Although there have been some notable successes, softwood timbers are still much preferred for pulp as well as for construction materials.

In the early 1970s, there was a growing acceptance of whole tree chips for pulp. Although a soft, worldwide pulp market since 1974 has curtailed the use of whole tree chips, their use is expected to expand. This will improve the opportunity to manage hardwood stands. Whole tree chipping nearly doubles the amount of wood recovered from hardwood stands where so much of the volume is in large limbs. This has stimulated research on thinning and other improvement cuttings of hardwood stands to increase their future value. Biltonen and others [13] discuss the need for a way to economically remove thinnings and defective trees to augment the supply of hardwood pulp chips and improve management of the 32 million acres of northern hardwood types in eastern United States. Their experiments in mechanized thinning of pole-size stands point the way to improved management of hardwood stands. Economical thinning methods allow recovery of wood otherwise lost to mortality and improve both the amount and value of wood available later for lumber and plywood.

A growing demand for wood fuel and a consequent increase in price could provide the needed incentive to stimulate management of forest lands in nonindustry, private ownership.

As pointed out by Dr. Zerbe's [14] paper in this volume, underutilized stands are not confined entirely to hardwoods. The Rocky Mountain area, for example, contains vast softwood resources that occur at long distances from market and pose challenging harvesting problems because of steep slopes and sensitive soils. In addition, underutilization of these stands may contribute to serious loss from disease and insects. Currently, an estimated 12 million cords of standing dead and down timber are accessible on National Forest lands of the Rocky Mountain region.

7. *Urban Residues*

A surprisingly large volume of wood is available as urban residue of various types. Tree trimmings, plus whole tree removals to combat Dutch elm disease or to make room for urban expansion, constitute substantial sources in many metropolitan areas.

An even larger amount of wood residue may be available in the form of demolition lumber, discarded pallets and crates, etc.

Auchter [7] has estimated that this source will provide 2 million cords/year by 1985 and 4 million tons per year by 2000 as a source of fiber. Knapp [15] has described experience in recovering newspaper, corrugated cartons, wood pallets, crating and demolition lumber from a mixed solid waste stream in one metropolitan area. Outputs of the system are again a mix of materials, including recycled newsprint, corrugated paper, pulp chips, particleboard furnish, and fuel. Knapp estimates that a metropolitan area of 1 million population, such as Portland, Oregon, will produce 1000 tons/day of wood and at least an equal amount of paper. Input is likely to be limited by the handling facilities and by available markets.

The potential of 50 million tons of reasonably dry wood residue from commercial waste will be of increasing interest to particleboard and pulp manufacturers, as well as to those seeking fuel, particularly as currently available residues become scarcer.

8. *Energy Plantations*

Solar radiation is the most abundant form of energy available. Unfortunately, the technology does not exist to utilize this source in significant amounts. Sunlight is diffuse and intermittent. Collection of feasible amounts of solar energy by current available means will require large land areas. Further, there is need to develop more practical ways of storing such energy.

Because of these problems, interest has developed in using energy plantations to accumulate biomass and subsequently convert this to heat or electricity. This interest in energy plantations has centered on trees because their experimental biomass yields of 4 to 25 tons per acre per year compare favorably with the production of such annuals as corn and sorghum, and the annual yields can be accumulated for 4 to 10 years.

Reported annual yields of biomass vary widely depending on species selection, spacing, land fertility, amount of nutrients and water applied, etc. A few representative yields are shown in Table VII, which is adapted from a 1974 report of the Stanford Research Institute [16]. These yields represent a solar energy conversion rate of about 0.2 to 1.2%, calculated by dividing the useful fuel value produced by the total solar radiation incident on the growing site during an entire year. A conversion rate of 0.4 to 0.7%, according to Kemp and Szego [17], is equivalent to the collection and storage, as above ground substance, of 80 to 140 million Btu's per acre per year.

In considering the intensive culture of trees for energy, it is well to note that forest plantations are already growing on nearly 200 million acres (80 million hectares), excluding China [1]; also, that many million acres of temperate hardwood stands in Europe, Australia, New Zealand, Chile, and Argentina have been converted to plantations of introduced softwoods [18].

The Nation's growing dependence on wood as a renewable natural resource has led to special programs of research aimed at providing the maximum yield of wood per acre. One such program is described in a report of five years research at the North Central Forest Experiment Station [19].

Brown [20] argues that intensive short-rotation management of forest plantations offers the best means of increasing wood fiber production and cites the following advantages: (a) higher yields per acre with smaller land requirements for a given production; (b) an earlier return on initial investment;

TABLE VII. Some Representative Yields Reported for Intensively Cultured Plantations[a]

Species	Location	Tons, oven-dry/ acre/yr[b]
Forage sorghum (irrigated)	New Mexico	7-10
Silage corn	Georgia	6-7
Sugar cane (10-yr average)	Hawaii	26
Sugar cane (5-yr average)	Louisiana	12.5
Hybrid poplar (short-rotation) stubble crop (3 years old)	Pennsylvania	8.7
American sycamore (short-rotation) seedlings (2 years old)	Georgia	4.1
Black cottonwood (2 years old)	Washington	4.5
Red alder (1-14 years old)	Washington	10
Eucalyptus spp	California	13.4-24.1

a. Adapted frpm Alich and Inman with permission [16].

b. 1 ton, oven-dry/acre-year = 4.30 m^3/hectare-year for wood density of 32.5 pounds per cubic foot. One m^3/hectare-year = 0.1786 cord/acre-year = 0.23 ton/acre-year.

(c) increased labor efficiency through mechanization of most operations, as in the intensified agriculture; (d) an opportunity to regenerate the stand by sprouts, thus reducing regeneration costs; (e) an opportunity to take advantage of cultural and genetic advances quickly. Among the disadvantages of short-rotation plantations are: (a) established management costs are higher than for conventional plantations; (b) only sites available to mechanized operations may be used; (c) short-rotation plantations would be large monocultures and therefore susceptible to epidemic disease and insect infestation.

Hardwoods, which sprout well, are best suited for short rotation management. The productive potential of hardwoods, however, can usually be achieved only on soils that are fertile, well drained, and adequately supplied with moisture throughout the growing season. Furthermore, a major forest management goal in the United States is to achieve a higher growth of softwoods and a higher use of hardwoods to better balance the domestic supply and demand.

Despite the appealing features of forest plantations as a source of clean energy, Calef [21] cites some serious shortcomings. He points out that regardless of the assumed biomass production rates, enormous amounts of land, water, and fertilizer would be required to supply even part of the Nation's energy needs with such plantations. Furthermore, he expresses the need to be aware of the disruption to life supporting systems that are provided by the natural growth of forests. Of special concern is potential soil deterioration through repeated harvest of the entire biomass and consequent depletion of humus and nutrients.

A further limitation on the establishment of energy plantations on desirable land is the fact that best annual returns (in $/acre) from forest plantations may be less than half the returns anticipated from corn or grain as revealed in Table VIII [22]. In addition, land suitable for energy plantations should have slopes of less than 25% and annual precipitation of over 20 inches. Much of the interest in biomass production has centered on lands of the Southwest, but availability of water is already a critical factor there.

Brown [20] argues that the enormous bulk of cellulose needed for the fiber markets of the world 25 years hence can only be produced in intensively managed, short-rotation forests. He bases his statement, in part, on the fact that we are unlikely to meet future fiber needs by improving trees on the basis of 25- to 30-year pulpwood rotations or still longer sawlog rotations. He further forecasts that tissue culture techniques will

TABLE VIII. Annual Net Returns From Short-Rotation Sycamore[a] *and Agricultural Crops for the Piedmont Area of Georgia* [22]

Crops	*Annual net return ($/acre)*
Bermuda hay	*117.80*
Oats and soybeans	*58.65*
Oats and grain sorghum	*45.96*
Soybeans	*44.24*
Cotton	*44.04*
Wheat	*35.11*
Sycamore (as an industrial enterprise receiving $10/ton, delivered)	*27.75*
Oats	*24.82*
Corn	*20.79*
Grain sorghum	*20.48*
Sycamore (as a farm crop returning $5/ton stumpage)	*8.13*

a. *Good site, 5-year cutting cycle.*

lead to rapid genetic improvement that can extend the effectiveness of short-rotation forestry.

It is my opinion that the major limitation on intensive culture short-rotation forestry--whether managed for fiber or fuel production or both--will be the competition for land, water, and fertilizer. Investments needed to insure adequate amounts of the three essentials may favor higher returns obtainable with other crops. Thus, forestry in general may be limited in the United States to those areas that already are in trees. Hardwood stands, in general, are producing far below their potential, especially on the 200 million acres of hardwood stands in farm and miscellaneous private ownership (40% of the Nation's commercial forest land). A thrust to improve forest productivity on these lands with their existing backlog of timber could pay large dividends. Removal cuttings to permit establishment of faster growing, more valuable species or partial cutting to upgrade the existing stands could provide millions of cords per year. For example, such management practices could recover a portion of the annual mortality loss of 15 million cords from the 1900 million cords of hardwood growing stock on nonindustry private lands.

Intensifying management on these lands not only avoids competition for the scarcer arable lands but also improves the productivity of lands already in trees. Granted, such factors as inefficient size of ownership, adverse local tax systems, and limited access to capital make such a program difficult. Creating forests from pasture or grasslands, however, must face the same difficulties to some degree.

V. COMPARATIVE VALUES OF WOOD FOR FUEL, FIBER, AND SOLID PRODUCTS

The value of wood for fuel is subject to two value comparisons. First, its value as fuel ranks low in comparison with its value as an industrial raw material (Table IX). Second, its value delivered to a furnace must be compared with available fossil fuels, particularly coal.

Corder [23] and Arola [24], have compared the heat values of various wood and bark fuels with those of fossil fuels. More meaningfully, they have compared their fuel values on a cost per million Btu recovered, considering the relative average efficiencies with which each is converted to steam. Arola's article includes a convenient chart to compare the values of solid fuels, such as wood, coal, and municipal waste, with those of gas or oil.

TABLE IX. Comparative Selling Values of Wood for Products and Fuel - 1976

	Approximate value per cubic foot
Export logs delivered to U.S. port at $330/M board feet	*$2.00*
Domestic logs delivered to mills at $100-250/M board feet	*0.65-1.65*
Export pulp chips $39/oven-dry ton (77 cubic feet)	*0.50*
Domestic pulp chips $35/oven-dry ton	*0.45*
Shavings $7.50/oven-dry ton	*0.10*
Sawdust $1.50-$5.00/oven-dry ton	*0.02-0.06*
Bark $1.00-$5.80/oven-dry ton	*0.01-0.07*

Because investment costs are higher for wood or coal firing than for gas or oil, the relative fuel values will be modified somewhat in favor of gas and oil. More important, the investment cost and payout period may determine when a company will make a change to solid fuel. Arola [24] has shown four methods of determining whether or not the investment in wood burning equipment is economically attractive. Tax incentives could influence this analysis and accelerate a changeover to wood or coal.

VI. PROJECTED INCREASES IN WOOD AND BARK USE FOR PRODUCTS AND ENERGY BY 2000

Seven sources of additional wood to fill increased U.S. needs for industrial raw material and energy are compared in Table X.

The surplus of timber growth over 1970 demand is expected to provide more than 50 million tons from the 70 million cords of excess growth estimated to be available in 2000.

Underutilized species are expected to be recovered from some 2 million acres a year. These can be expected to produce an average of 25 cords per acre, as stands of low value are converted to a more valuable species mixture. Increasing competition should raise wood prices and foster interest in managing underutilized stands to supply an estimated 60 million tons of wood and bark annually by 2000.

Thinnings or other improvement cuttings are anticipated on at least 2 million acres per year, with an average removal of 10 cords per acre. These cuttings should supply another 23 million tons for products and energy. This outlet will also help improve the management of forest lands.

Logging residue from live tree removals is expected to provide about 14 million cords of additional material in 2000. In the West, particularly, there are additional residues from dead or down trees to augment tops and broken stems of live trees. Overall, logging and forest residues are expected to provide 20 million cords (26 million tons) of wood and bark in 2000.

Urban wastes, including demolition lumber, crates, pallets, etc., are recoverable in most metropolitan areas, but values have not encouraged recovery until now. It is anticipated that, because of their proximity to markets, 50% of the 50 million tons available may be used by 2000.

Energy plantations are expected to contribute only a limited amount of material in 2000. Costs and institutional

TABLE X. *Projected Increases in Wood and Bark Use for Products and Energy by Source (Additions to 1970 Use)*

Source of wood	Additional quantity estimated to be available	Projected use of additional supplies in 2000	
		For products	For fuel
	(millions of short tons, oven-dry)		
Manufacturing residues	29	12	17
Surplus roundwood growth			
Softwood[a]	22	10	8
Hardwood[b]	65	6	25-30
Underutilized species[c]	60	-	50
Thinnings, improvement cuttings[d]	23	15	8
Logging residues	26	17	9
Urban wastes	50	10	18
Energy plantations	(20)	(10)	(10)
Totals	275	70	135-140

a. 140 million cords growth less 120 million cords used in 1970 × 1.1 ton/cord.
b. 105 million cords growth less 55 million cords used in 1970 × 1.3 ton/cord.
c. Stand conversion, etc. on 2 million acres per year at 25 cords per acre × 1.2 tons/cord.
d. Cuttings on 2 million acres per year at 10 cords/acre × 1.15 ton/cord.
e. Energy plantations conceivably could provide wood from 2 million acres at 10 tons/acre/year but are considered the least likely source and are not included in the totals.

problems faced in consolidating suitable land areas and in insuring adequate water are likely to limit this endeavor to perhaps 2 million acres of production by 2000, although intensive culture of short-rotation forest plantations could continue to increase after that time. A potential contribution of 20 million tons, oven-dry, from this source is anticipated in 2000.

VII. CONCLUSIONS

Overall needs of the forest products industries for additional wood and bark by the year 2000 are estimated to be about 70 million tons, oven-dry, for products and over 140 million tons, oven-dry, of fuel for steam production.

The seven potential sources of additional wood and bark considered herein can supply more than the above requirements as shown in Table X. It becomes a matter of opinion concerning the relative quantities and specific end uses to be supplied by each source. No one source is likely to provide its full potential by the year 2000. Geographical distribution of potential users and potential suppliers, as well as institutional and technical limitations, will require time to resolve.

For the sake of discussion, however, a plausible distribution of materials from several sources to the key segments of the forest products industries--all based on the estimated needs of these industries by the year 2000--is summarized in Table X.

The distribution of additional supplies concentrates on needs of the forest products industries to the exclusion of other potential users. This is done for four primary reasons:

1. The pulp and paper industry has massive needs for additional wood, both for anticipated fiber needs and for wood and bark fuels to guard against shutdowns because of petroleum fuel shortages.

2. The forest products industries have the best opportunity to use available but underutilized wood resources because their plants are generally close to potential supplies, and they are experienced in handling bulk, wet, and often dirty materials.

3. The forest products industries can generate energy from wood and bark more efficiently than others because of their heavy process steam or heat requirements compared with their electric power needs. Use of low pressure exhaust steam for process drying, pressing, etc., can increase turbine-electric generating efficiency from near 25% (when generating steam from wood) to 50 or 60% efficiency. The paper by Dr. Johanson and Dr. Sarkanen presents a detailed examination of this approach.

4. Every ton (oven-dry basis) of wood or bark burned in the forest products industries releases an average of 2.4 barrels of oil or equivalent. The potential saving is equivalent to more than 300 million barrels of oil per year based on projected increases in wood fuel use by the forest products industries (Table X). This is equivalent to about three fourths of the oil expected annually from Alaska's North Slope.

The additional wood that will be supplied to a particular segment of the forest products industries, to other industries, or to public utilities will depend on a great many factors. The additional supplies and uses of wood forecast here can be changed by circumstances.

REFERENCES

1. Risto Eklund, "The Future of Wood as a Renewable Raw Material." Paper presented at Inauguration of the Jaakko Poyry & Co. Headquarters in Helsinki, September 12, 1975.
2. David A. Tillman, "The contribution of Non-Fossil Organic Materials to U.S. Energy Supply." Federal Energy Administration Contract No. P-03-77-4426-0, February 1977.
3. Stephen H. Spurr and Henry J. Vaux, "Timber: Biological and Economic Potential." Science 191 (4228), 1976.
4. Forest Service, U.S. Department of Agriculture, "The Outlook for Timber in the United States." For. Resour. Rep. No. 20, Washington, D.C., 1973.
5. S. H. Schurr and Bruce C. Netschert, Energy and the American Economy, 1850-1975. John Hopkins Press, 1960.
6. Conor W. Boyd, Peter Koch, Herbert B. McKean, Charles R. Morschauser, Stephen B. Preston, and Frederick F. Wangaard, "Wood for Structural and Architectural Purposes." Special CORRIM Panel report, Wood and Fiber 8(1), 1976
7. Richard J. Auchter, "Raw Material Supply" Future Technical Needs and Trends in the Paper Industry-II, Committee assignment report (CAR) No. 64 (TAPPI), Tech. Assoc. Pulp and Paper Industry, 1976.
8. Robert W. Hagemeyer, "Future World Demand for Paper and Paperboard--and the Geography of the Technical Needs," Future Technical Needs and Trends (CAR) No. 64, (TAPPI), 1976.
9. Ronald J. Slinn, "Some Aspects of Energy Use by the U.S. Pulp and Paper Industry." American Paper Institute, 1974.
10. Forence K. Ruderman, "Production, Price, Employment, and Trade in Northwest Forest Industries," First Qtr. USDA For. Serv., Pacific Northwest For. Exp. Stn, Portland, Oregon, 1976.

11. John B. Grantham and Thomas H. Ellis, "Potentials of Wood for Producing Energy." J. For. 72:552, September 1974.

12. Thomas C. Adams. "Chipmill Economics Eyed by Northwest Industry." For. Industries, June 1977.

13. Frank E. Biltonen, William A. Hillstrom, Helmuth M. Steinhilb, and Richard M. Godman, "Mechanized Thinning of Northern Hardood Pole Stands: Methods and Economics." USDA For. Serv. Res. Pap. NC-137, North Cent. For. Exp. Stn., St. Paul, Minn., 1976.

14. John I. Zerbe, "Conversion of Stagnated Timber Stands to Productive Sites and Use of Noncommercial Material for Fuel." For presentation to Cellulose and Fuels Division, American Chemical Society, 1977.

15. John H. Knapp, "Potential of Industrial Wood Residue for Energy." Proceedings of the FPRS conference on Energy and the Wood Products Industry, Atlanta, Ga., November 1976.

16. Stanford Research Institute, "Effective Utilization of Solar Energy to Produce Clean Fuel," 1974.

17. Clinton C. Kemp and George C. Szego, "The Energy Plantation." Symposium on Energy Storage, 168th American Chemical Society National Meeting, Atlantic City, N.J. 1974.

18. J. S. Bethel and G. F. Schreuder, "Forest Resources: An Overview." Science 191 (4228), 1976.

19. Forest Service, U.S. Department of Agriculture. "Intensive Plantation Culture: Five Years Research." USDA For. Serv. Gen. Tech. Rep. NC-21, North Cent. For. Exp. Stn., St. Paul, Minn., 1976.

20. Claud L. Brown, "Forests As Energy Sources in the Year 2000: What Man Can Imagine, Man Can Do." J. For. 74(1), 1976.

21. Charles E. Calef, "Not Out of the Woods." Environment, September 1976.

22. G. F. Dutrow and J. R. Saucier, "Economics of Short-Rotation Sycamore." USDA For. Serv. Res. Pap. SO-114, South. For. Exp. Stn., New Orleans, La., 1976.

23. Stanley E. Corder, "Fuel Characteristics of Wood and Bark and Factors Affecting Heat Recovery." Proceedings of FPRS Conference on Wood Residue as an Energy Source, Denver, Colo., 1975.

24. Rodger A. Arola, "Wood Fuels--How Do They Stack Up?" Proceedings of FPRS Conference on Energy and the Wood Products Industry, November 15-17, Atlanta, Georgia, 1976.

THERMAL ANALYSIS OF FOREST FUELS

Fred Shafizadeh and William F. DeGroot

Wood Chemistry Laboratory
University of Montana
Missoula, Montana

I. INTRODUCTION

Forest fuels and biomass in general are composed of cellulose, lignin and a variety of other components such as extractives, water, and ash. The relative quantities of these materials affect, not only the total heat of combustion of the fuels, but also the net energy which is released and the rate of energy release under different conditions of pyrolysis and combustion [1-3]. At elevated temperatures pyrolysis of biomass produces a carbonaceous char and a variety of volatile degradation products, some of which could condense to form a liquid or tar fraction. The relative proportions of these products are dependent not only on the composition of the fuel, but also on the reaction temperature, rate of heating and availability of oxygen, which together dictate the pyrolysis or combustion pathway followed. Ignition of the volatile pyrolysis products in air results in flaming combustion and rapid evolution of heat, while the carbonaceous char burns in the solid phase by surface oxidation, which is a slower process.

Thus, composition of the fuel and reaction conditions exert a strong influence on the thermal reaction pathways and rate of combustion or heat release. The competing pyrolysis and combustion reactions provide a means of converting biomass to different types of solid, liquid or gaseous fuel, with or without production of chemicals that could be isolated from the pyrolysis products. Thermal analysis provides a useful method for analyzing the sequence of reactions and the accompanying energy and mass transformations which take place as the substrate is heated.

II. THERMAL ANALYSIS METHODS

The most widely used methods of thermal analysis are thermogravimetry (TG), differential scanning calorimetry (DSC), and differential thermal analysis (DTA). Thermogravimetry provides a continuous recording of the weight or mass transformation of a sample as a function of temperature or time, as shown in Fig. 1. Derivative thermogravimetry (DTG), which is usually determined simultaneously, provides the rate of weight loss under the same heating conditions. These methods have been used extensively for characterizing fuels and determining the kinetics of their gasification.

The temperature and enthalpy changes associated with chemical or physical transformations can be determined by differential scanning calorimetry (DSC). A closely related method, differential thermal analysis (DTA) provides similar qualitative rather than quantitative information. As shown in Fig. 2, the DSC signal can be integrated to give the total enthalpy change per unit weight of fuel within the selected temperature interval.

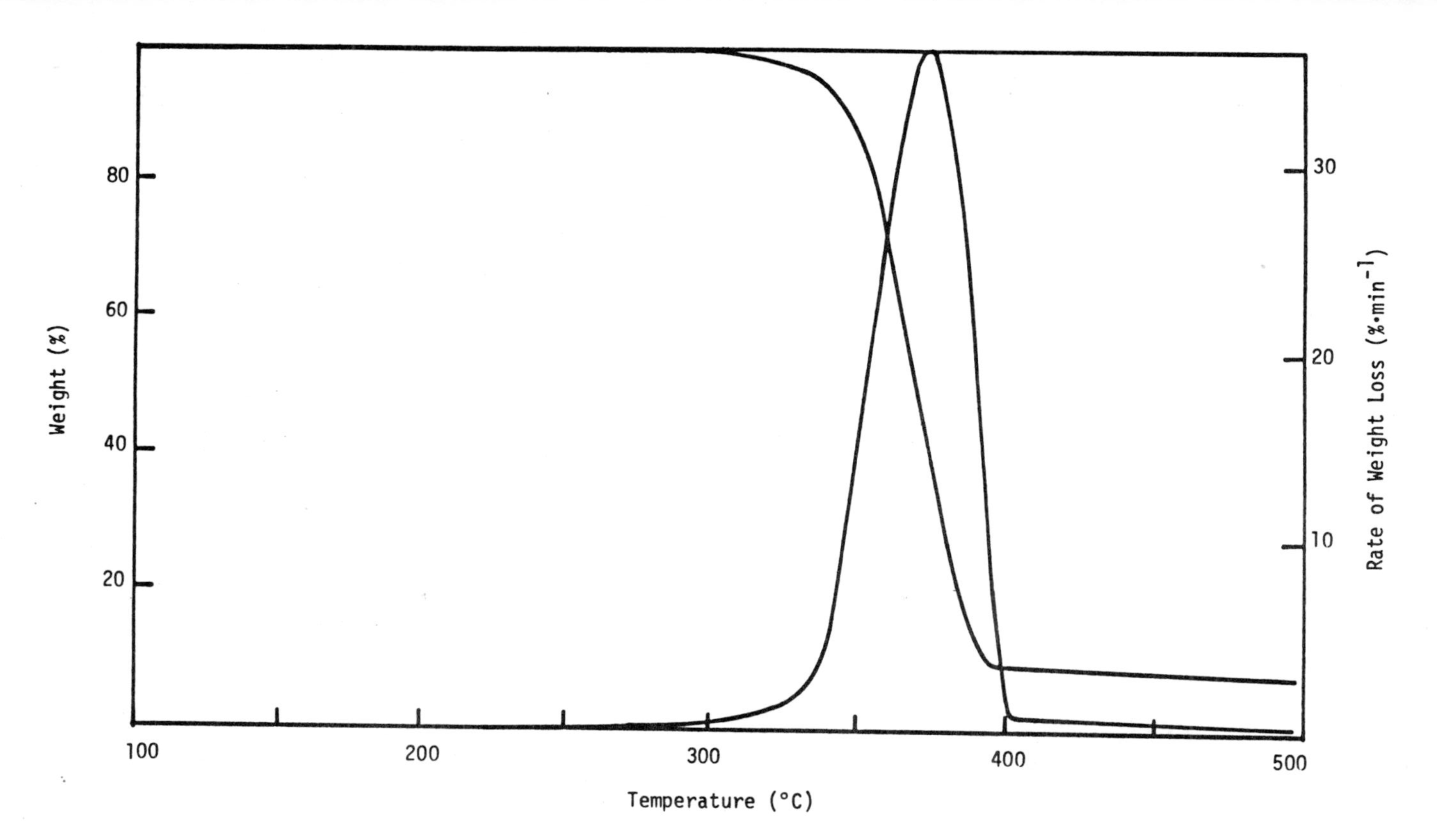

FIGURE 1. Thermogravimetry of cellulose (heating rate 15°/min).

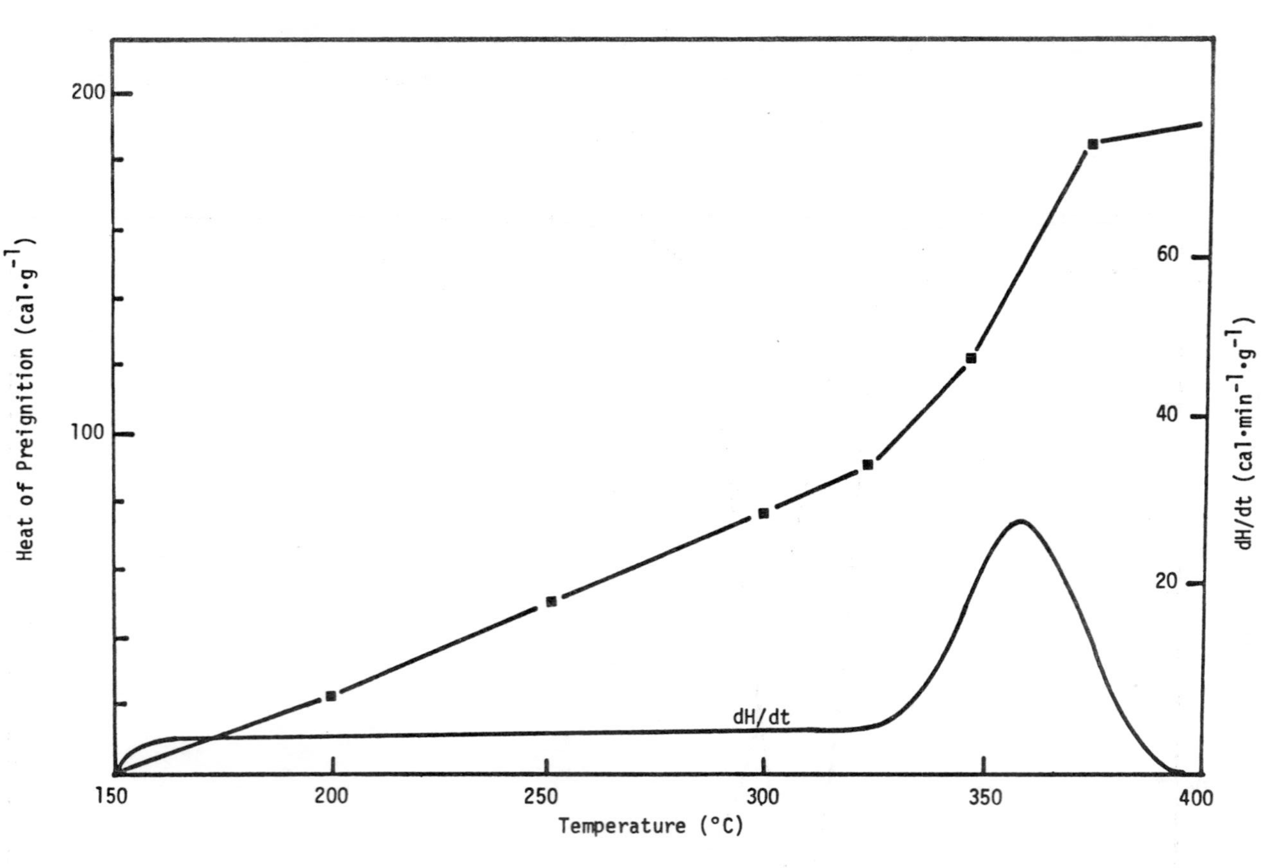

FIGURE 2. Differential scanning calorimetry of cellulose (heating rate 10°/min).

A more recently developed method, thermal evolution analysis (TEA), involves the use of an appropriate detector to determine the temperature at which volatile degradation products are formed. The reaction coulometer detector [4] has recently been adapted for use with this method, extending its capabilities to the measurement of the heat of combustion of evolved gases. This allows for quantitative determination of the heat of combustion of volatile pyrolysis products and significantly enhances the ability to model combustion or pyrolysis processes.

This TEA system, illustrated in Fig. 3, uses a closed-loop system of detection involving an electrolytic oxygen generator and detector and a reactor where gases evolved by the heated sample undergo combustion. A constant level of oxygen is fed into the reactor during the run; as volatiles are produced, their combustion causes a deficit in the oxygen level which is exactly offset by increased generation of oxygen. The voltage applied to the oxygen generator is recorded throughout the process. The recorder output is thus equivalent to the oxygen consumption required for complete combustion of the evolved gases. This oxygen requirement is a measure of the extent of the combustion reaction at any time and can be related to the rate of heat release or integrated to give the cumulative heat of combustion of evolved gases, the "effective heat content," for a fuel heated to any temperature. Comparison of heats of combustion of pure compounds reported in the literature with the oxygen requirement for their complete combustion indicates that the conversion from oxygen requirement, measured by the reaction coulometer, to heats of combustion can be made with about 93% precision. The TG and TEA methods thus complement each other in that TG gives the mass of the solid and, by difference, gaseous fractions in pyrolysis; while TEA gives the energy content of the volatiles and, by difference, the energy content of the charred residue. The ratio of the values obtained by the two methods provides an index of the

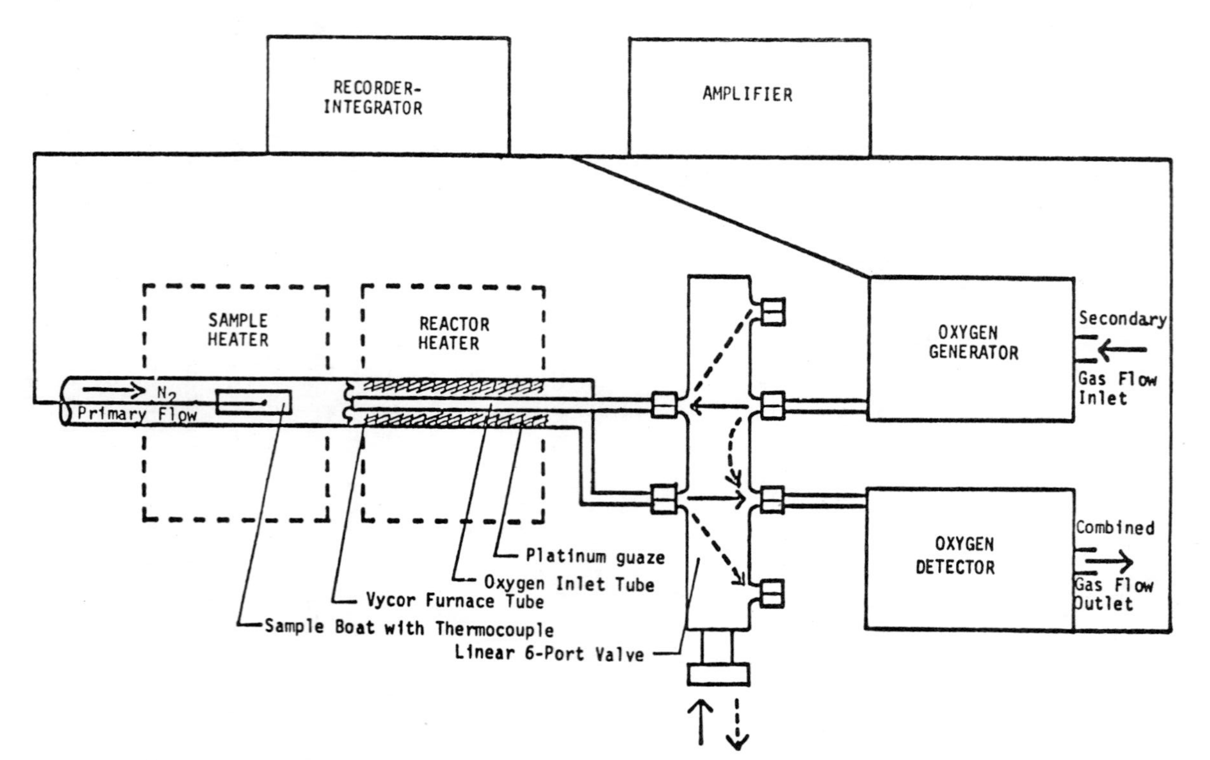

FIGURE 3. Block diagram of the thermal evolution analysis system and reaction coulometer. (Single lines represent electrical connections; double lines represent gas connections.)

concentration of combustible gases in the volatile fraction. This concentration index, which is a measure of the potential of a fuel to generate an ignitible fuel: air mixture, provides the basis for the "flammability index" discussed in a previous paper [5].

The TEA curve in Fig. 4 shows the rate of oxygen consumption in combustion of the pyrolysis products of cellulose, and the integral of this rate curve, which gives the cumulative oxygen consumption. The corresponding rate of release of combustion heat is also indicated.

The complex natural fuels present a unique problem for thermal studies since they are composed of many components having distinct thermal characteristics. These fuels can be analyzed as a whole, or the individual components can be isolated and analyzed individually. As shown in Fig. 5, the thermal response of the whole fuel is simply the sum of the thermal responses of its components [6], ensuring applicability of analysis of components to the thermal response of the whole fuel. Analysis of components is often advantageous, because it provides more detailed information on the individual reactions which contribute to the thermal degradation of the fuel and also allows for estimation of the thermal properties of a fuel based on its chemical composition.

III. ANALYSIS OF COMBUSTION PROCESSES

The simplest reaction system requiring definition of thermal parameters is one involving direct combustion. Provided the combustion is carried out at a sufficiently high temperature and concentration of oxygen, the entire heat content of the fuel will be liberated, producing carbon dioxide and water. At elevated temperature, however, the heat of combustion determined at room temperature must be corrected for the enthalpies of the reactants

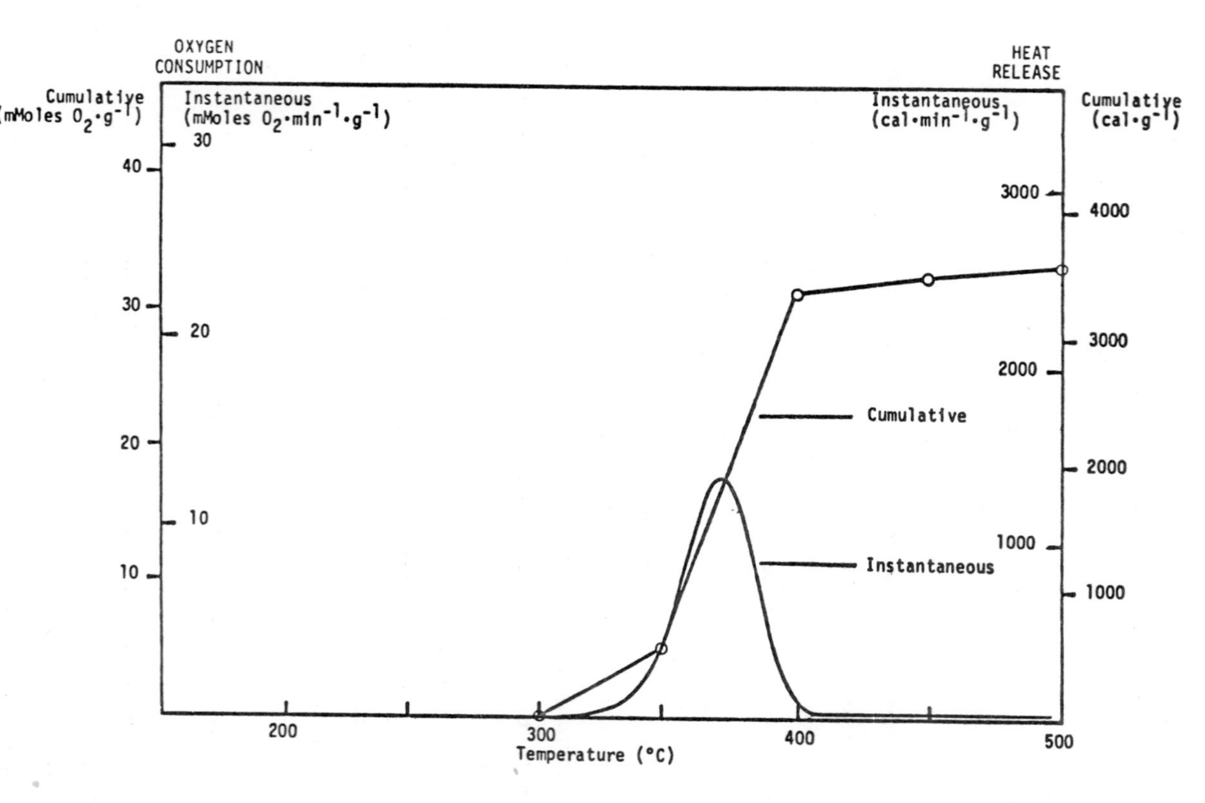

FIGURE 4. *Thermal evolution analysis (TEA) of cellulose (heating rate 15°/min).*

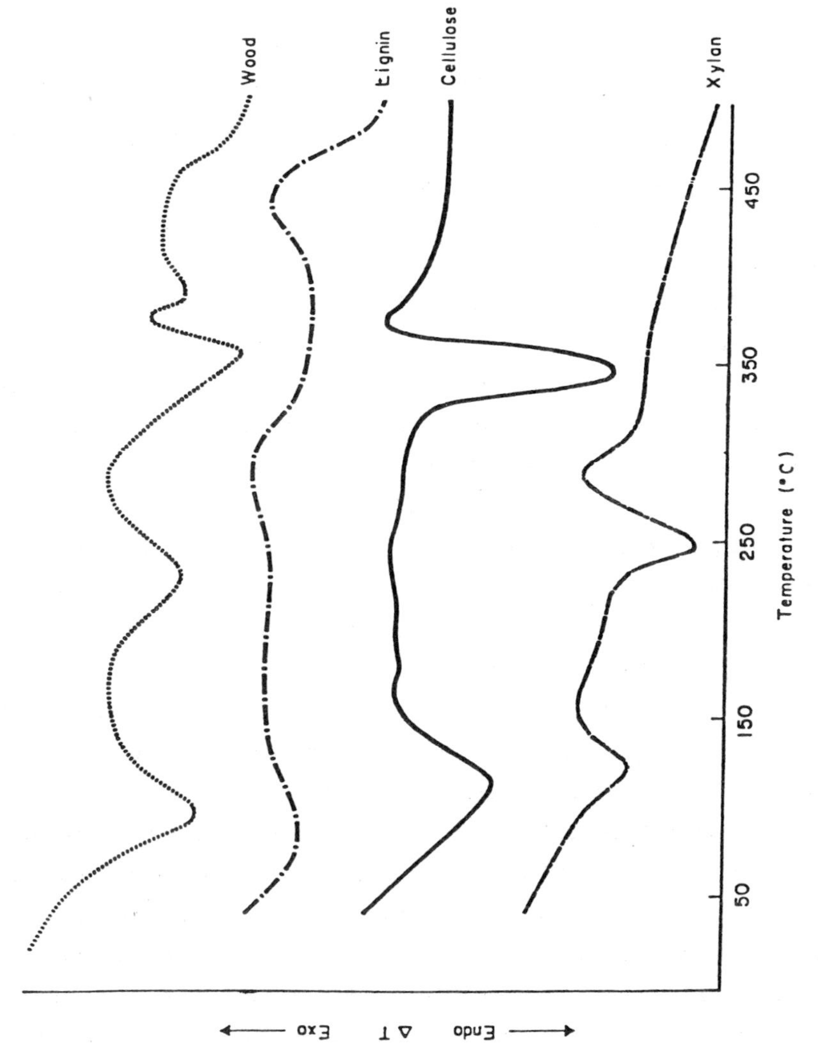

FIGURE 5. Differential thermal analysis of wood and its components.

and products. The heat of combustion at these temperatures will be lower than that determined at room temperature, and can be calculated for the reaction, fuel + $O_2 \longrightarrow CO_2 + H_2O$, according to the following equation [1]:

$$\Delta H^{x°}_{comb} = \Delta H^{25°}_{comb} + \int^{x°}_{25°} Cp\ (CO_2)dT + \int^{x°}_{25°} Cp\ (H_2O)dT$$
$$+ \Delta H^{25°}_{vap}\ (H_2O) - \int^{x°}_{25°} Cp\ (fuel)dT - \int^{x°}_{25°} Cp\ (O_2)dT \quad (1)$$

The heat of vaporization term and heat capacity terms for gases are available from the literature, the latter by integrating the heat capacity equations of Spencer [7]. The quantities of O_2 required and CO_2 and H_2O produced in this process can be determined from elemental analysis of the fuel. The only term in the expression which must be experimentally determined is the heat capacity of the fuel, which is determined by DSC as described previously and illustrated in Fig. 2.

Experimentally determined values of enthalpy changes of several natural fuels heated from 25° to 400° and corresponding calculated values for reactant and product gases are given in Table I [8]. The values given for the fuels are the sum of the enthalpy changes due to heat capacity of the fuel, heat of pyrolysis, heat of vaporization of product gases and heat capacity of the char over the temperature ranges given.

Measured and calculated heats of combustion of the natural fuels are given in Table II. The heats of combustion at 400° are calculated from Equation 1. The correction for temperature is small compared to the total heat of combustion, although certainly not negligible. The dominant factor in the calculation is the heat of vaporization of the water produced by combustion. This accounts for the larger correction necessary for cellulose, which has a relatively high hydrogen content, and indicates that the correction will decrease for fuels such as lignin, which has a high degree of unsaturation in its molecular structure. Since the

TABLE I. Heat Required to Raise Temperature of Fuels and Combustion Gases from 25° to 400°.

Sample	$\Delta H_{25°-400°}$ (cal/g)
Cellulose	179.0
Wood (Populus spp.)	190.0
Punky wood (Douglas fir)	201.0
Ponderosa pine needles	230.0
Oxygen	87.0
Carbon dioxide	89.2
Water	175.0

TABLE II. Heats of Combustion of Natural Fuels at 25° and 400°.

Sample	$\Delta H^{25°}_{comb}$ (cal/g)	$\Delta H^{400°}_{comb}$ (cal/g)	Correction 25°-400° (cal/g)
Cellulose	-4143	-3853	290
Wood (Populus ssp.)	-4616	-4341	277
Punky wood (douglas fir)	-5120	-4878	242
Ponderosa pine needles	-5145	-4904	241

contribution of the heat capacity of product gases is small ompared to the heat of vaporization term in Equation 1, the heat of combustion correction is not expected to vary significantly with temperature. The heat capacity and latent heat corrections in Equation 1 apply only to the adjustment of the heat of combustion values, however, and allowance must still be made for the enthalpy changes of the fuel determined by DSC in determining the energy balance in a thermal conversion system. Therefore, the sensible and latent heat requirements of the fuel will demand greater thermal energy for processes carried out at higher temperatures, although the correction of the heat of combustion will not change significantly from that at lower temperatures.

IV. ANALYSIS OF PYROLYSIS SYSTEMS

In analyzing the energy balance in a pyrolysis system, it is necessary to know (1) the energy required to bring the fuel to the reaction temperature, or "heat of preignition" determined by DSC, (2) the fraction of the heat content of the fuel available in the gas phase, or "effective heat content" determined by TEA, and (3) the fraction of the heat content remaining in the solid phase, determined by difference of the effective heat content and heat of combustion of the fuel. It is also necessary to know the quantity of volatiles and residues remaining at any temperature, determined by TG, for conversion of the above values to a weight basis.

The calculation of the heat required to bring the fuel to ignition was discussed in the previous section. The effective heat content determination has been made by various methods. The most fundamental expression of heat release or reaction intensity, I_R, is:

$$I_R = dw/dt\ (h)$$

where dw/dt is the rate of weight loss determined by DTG and h is the heat of combustion of the fuel. This provides only an approximation of the actual heat release, however, in that the heat of combustion of the solid is not constant over the duration of the reaction, but is increasing as the sample becomes more highly carbonized.

The original TEA work conducted on forest fuels at the Wood Chemistry Laboratory involved detection of combustible volatiles by flame ionization [5,9]. While this method provided accurate information on the temperature dependence of the evolution of combustible volatiles, quantitative data could be obtained only by calibration of the FID response with compounds of approximately the same oxidation level as those encountered in the volatiles. This method provides reasonable accuracy for pure compounds, but for complex fuels such as plant biomass, it is best applied to determination of relative, rather than absolute values.

The adaptation of the reaction coulometer detector to the TEA system represents the first quantitative method of determining effective heat content. A representative TEA curve of fresh Douglas fir needles is given in Fig. 6, along with the TG of the same fuel. Foliage biomass provides a good example of the application of the TEA method, since it is composed of appreciable levels of a broad spectrum of chemical constituents including extractives, carbohydrates, phenolics and adsorbed water. The TEA curve reflects the combined thermal response of the combustible fraction of the fuel, but is insensitive to the evolution of noncombustible gases, such as the large quantity of water evident in the TG curve.

The pyrolysis and evaporation of the extractives begins about 150° and increases slowly until the physical structure of the needle begins to break down at about 200°. The evolution of highly volatile oils or accumulated pyrolysis products accompanying this loss of physical integrity of the fuel gives rise to the sharp peaks above 200°. The most rapid rate of evolution, which occurs at a temperature of 350°, corresponds to the degradation of the carbohydrate and phenolic fractions. The evolution of combustible volatiles at higher temperatures is due to the slow decomposition of the carbonized extractive, carbohydrate and phenolic fractions. The effective heat content at 500° of -2730 cal/g amounts to 53% of the initial heat of combustion of

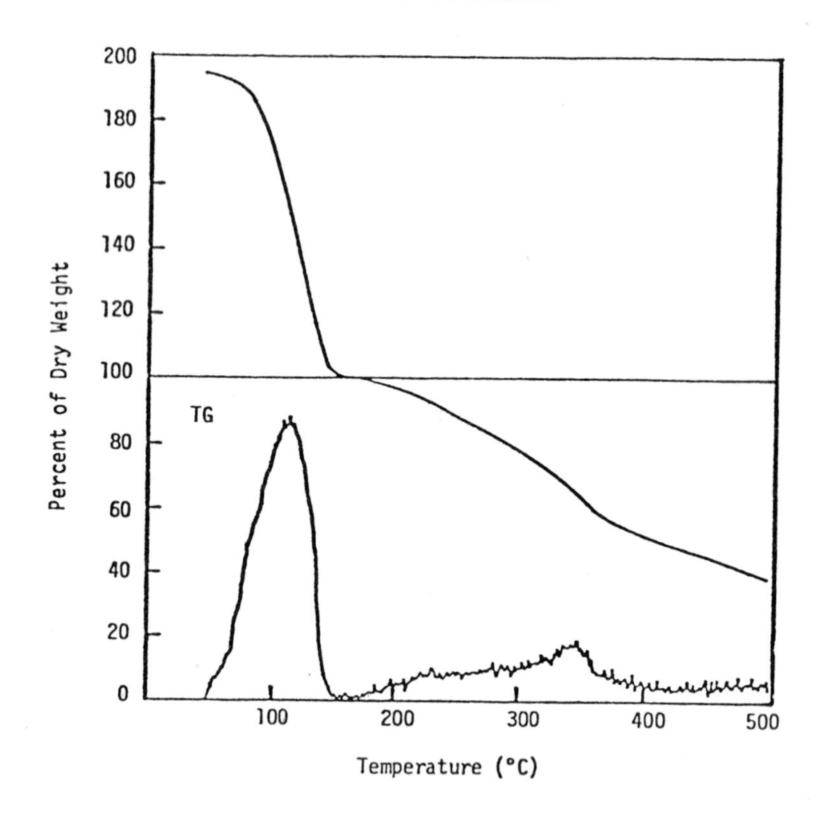

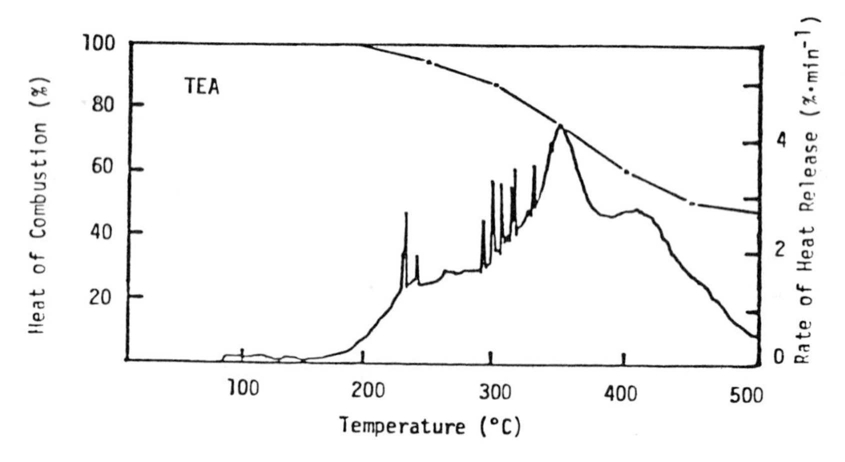

FIGURE 6. Weight loss (TG), heat release and oxygen consumed in combustion (TEA) for fresh Douglas fir foliage.

the fuel. The TG curve of the same fuel, also shown in Fig. 6, indicates a residual weight over the same temperature range of 35%. From the combined data from these two methods, the individual heats of combustion can be calculated on a weight basis for the char and volatile fractions, as shown in Table III.

Representative TG and TEA curves for cotton wood heartwood (*Populus trichocharpa)*, cottonwood Klason lignin and cellulose are shown in Figs. 7-8. The figures indicate the degree of volatilization of the fuel and the distribution of the heat content of the fuel between the solid and volatile fractions.

TABLE III. Distribution of the heat of combustion of Douglas fir foliage during pyrolysis.

Temp (°C)	*Weight loss (%)*	*Distribution of heat of combustion (cal/g fuel)*		*Distribution of heat of combustion (cal/g)*	
		Volatiles	*Char*	*Volatiles*	*Char*
100	-	*13*	*5120*	*(--)*	*5120*
150	-	*26*	*5107*	*(--)*	*5107*
200	*4.5*	*60*	*5073*	*1333*	*5310*
250	*13.5*	*297*	*4836*	*2200*	*5590*
300	*24.6*	*644*	*4489*	*2620*	*5950*
350	*40.0*	*1268*	*3865*	*3170*	*6440*
400	*50.0*	*1992*	*3141*	*3980*	*6280*
450	*56.3*	*2500*	*2633*	*4440*	*6030*
500	*63.5*	*2730*	*2403*	*4300*	*6580*

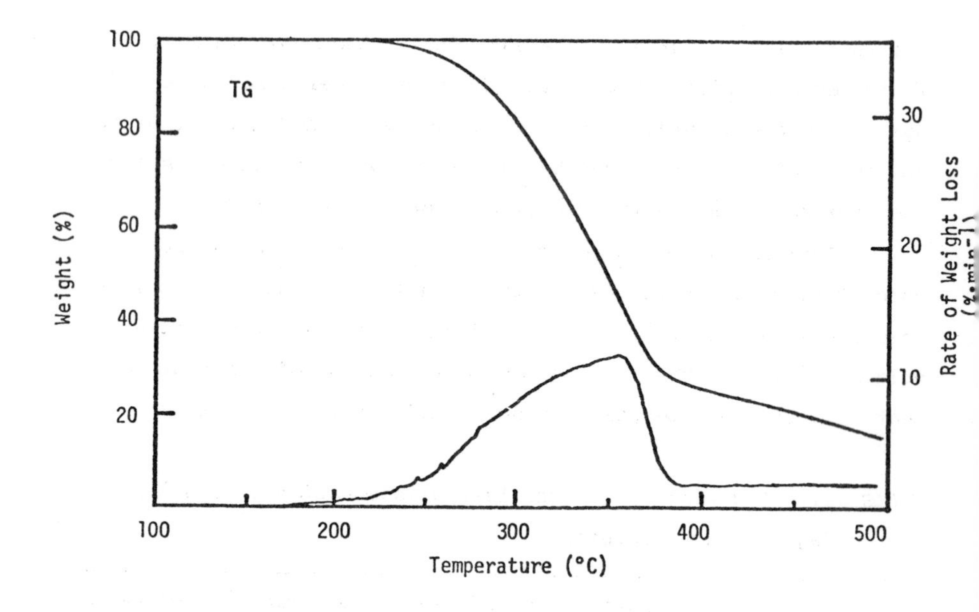

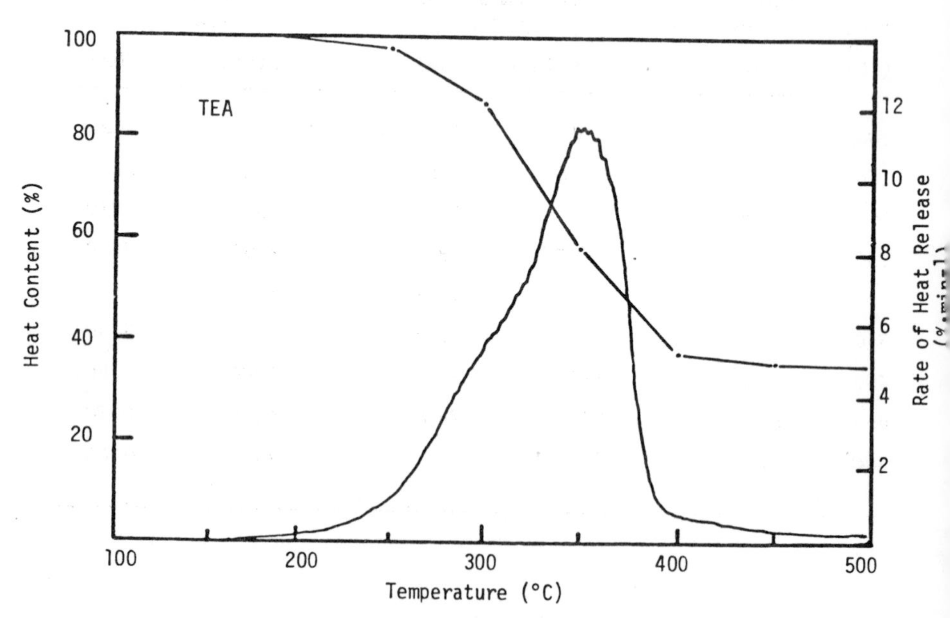

FIGURE 7. Weight loss (TG) and heat release (TEA) from ground cottonwood, ΔHcomb-4618 cal/g (heating rate, 15°/min).

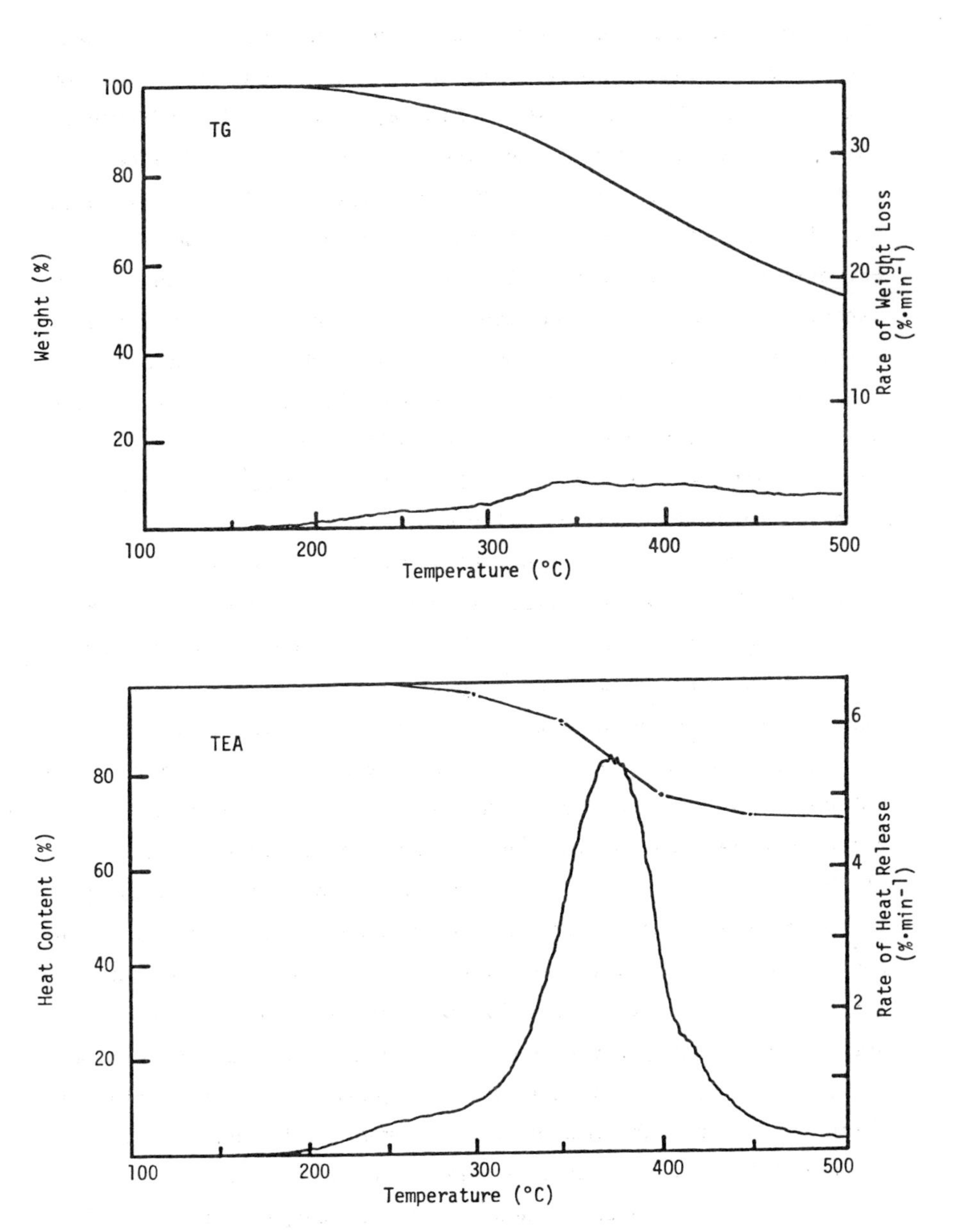

FIGURE 8. Weight loss (TG) and heat release (TEA) curves for cottonwood Klason lignin, ΔHcomb-6370 cal/g (heating rate 15°/min).

The thermal analysis of cottonwood, given in Fig. 7, shows that 84% of its weight, but only 65% of its heat content is lost on pyrolysis to 500°. As shown in Fig. 8, most of this difference is due to the charring of the lignin in which 70% of the heat content remains in the char. Thermal analysis of the cellulosic fraction, shown in Fig. 9, indicates that 92% of the weight and 86% of the heat content is lost as volatiles. Thus, the lignin, although its heat of combustion is much higher than that of cellulose, does not contribute significantly to the heat content of the gases, since it predominantly chars with most of its combustible carbon content remaining in the solid phase. This results in an effective heat content of less than 2000 cal/g for lignin, compared to 3500 cal/g for cellulose.

The combination of the TEA and DSC methods can be used to determine the net energy yield from a pyrolytic system as shown in Fig. 10 for cellulose. In this figure, the energy required to heat the fuel at a given temperature, the "heat of preignition," calculated by integration of the DSC curve, is compared with the effective heat content, calculated by integration of the TEA curve for wet and dry cellulose. At lower temperatures, there is a net heat deficit due to the heat capacity requirements of the fuel. This deficit is very quickly offset, however, by the effective heat content as the fuel begins to decompose. The net energy yield at any temperature can be calculated as the difference between the effective heat content and the heat of preignition. The point at which the curves cross is the minimum ignition temperature in fire modeling because it is the point at which the overall thermal balance becomes exothermic. It is evident from Fig. 10 that adsorbed moisture can have a significant effect on the energy yield, while the temperature of ignition is less dependent on the moisture content. The net energy yield from the dry fuel is approximately 600 cal/g (dry weight) greater than that of the wet fuel, the difference being due to the heat of vaporization of adsorbed water.

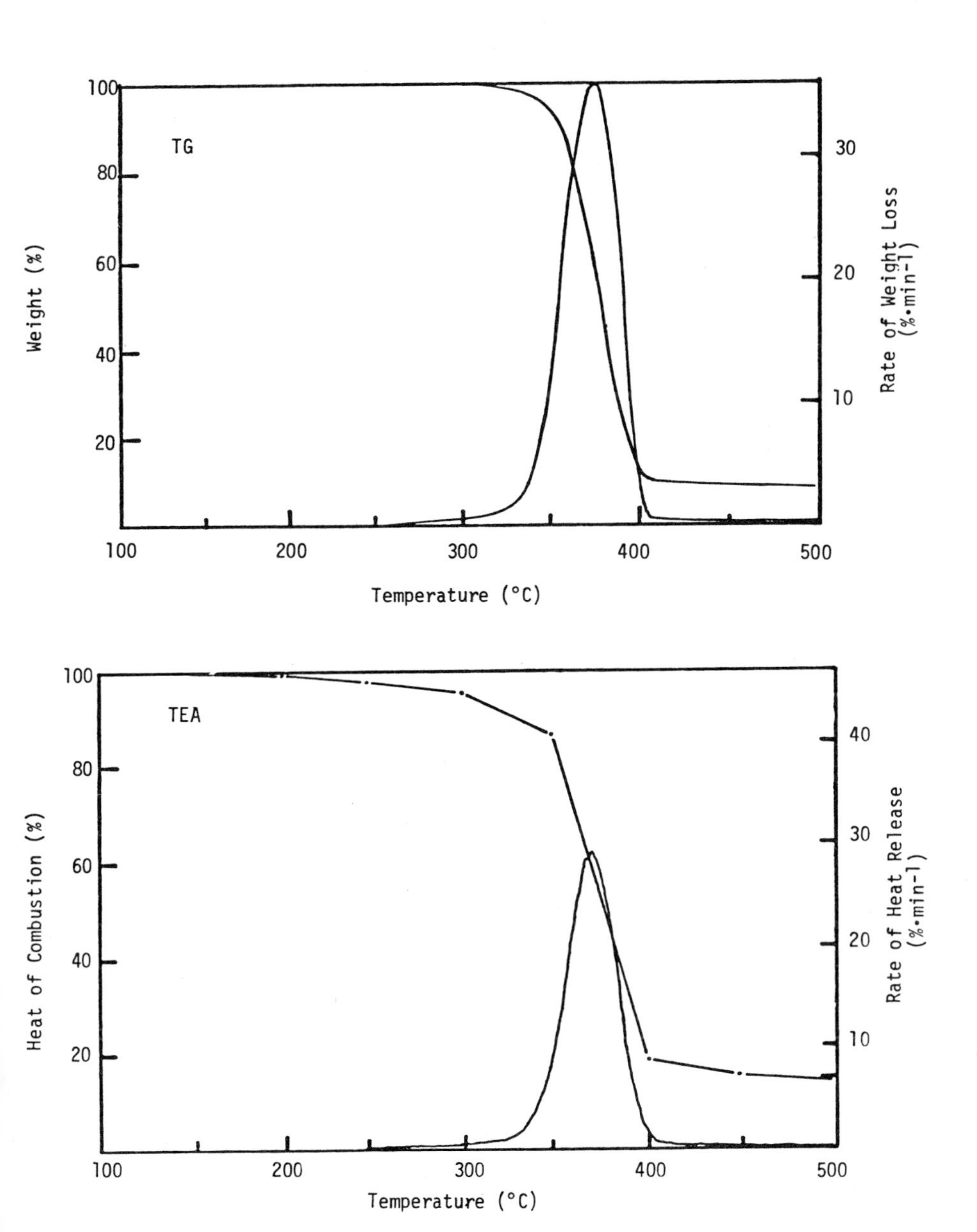

FIGURE 9. Weight loss (TG) and heat release (TEA) from cellulose, ΔHcomb-4143 cal/g (heating rate 15°/min).

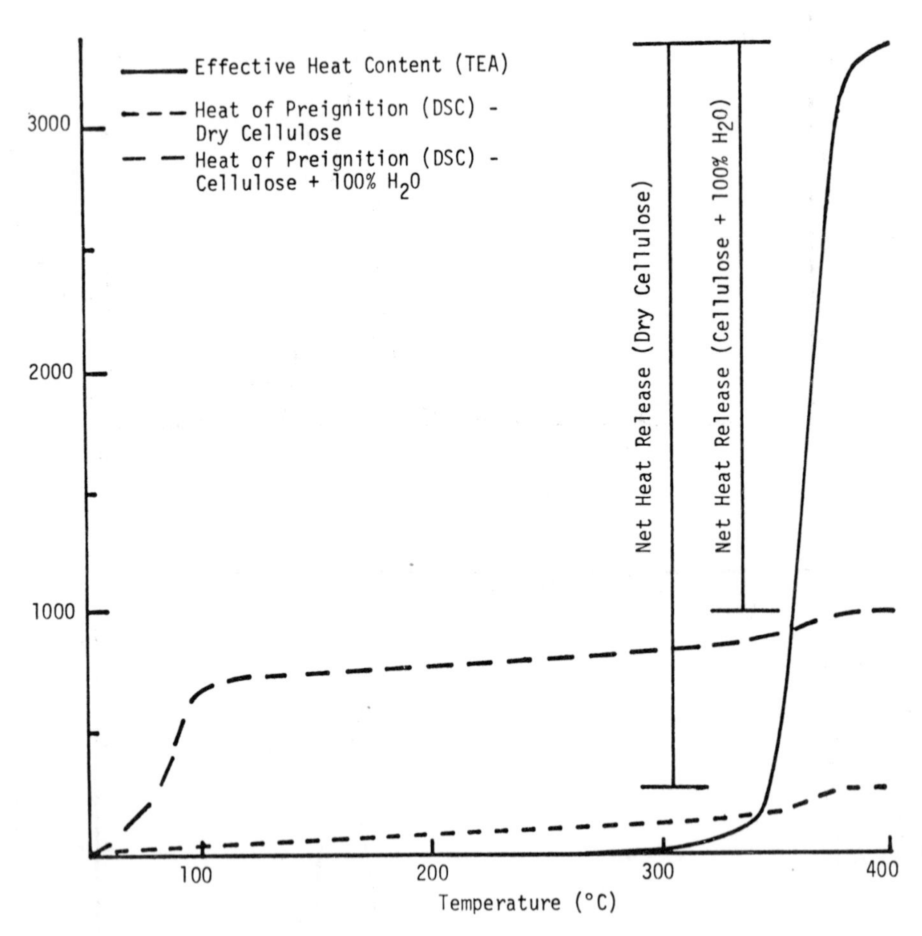

FIGURE 10. Comparisons of heats of preignition and effective heat content of dry and wet cellulose.

V. CONCLUSIONS

With the recent development of a quantitative method of TEA, current techniques of thermal analysis can provide the data necessary for accurate evaluation of material and energy balances in pyrolysis or combustion. The net energy available from combustion processes is reduced by 250-300 cal/g in correcting the heat of combustion to the reaction temperature and by a similar quantity in accounting for the heat of preignition of the fuel.

In pyrolysis, the energy content of the fuel is distributed to varying degrees between the char and volatile fractions depending on the composition of the fuel and the reaction conditions employed. Volatilization is enhanced by high carbohydrate and extractive content in the fuel and elevated reaction temperatures. Thermal analysis provides the fundamental data necessary for evaluation and characterization of biomass as fuel and formulation of thermal conversion processes.

ACKNOWLEDGMENT

The authors are pleased to acknowledge the support of The National Fire Prevention and Control Administration and the U.S. Forest Service for this study.

REFERENCES

1. R. A. Susott, W. F. DeGroot and F. Shafizadeh, J. Fire & Flammability, 6, 311, (1975).
2. F. Shafizadeh and W. F. DeGroot in Thermal Uses and Properties of Carbohydrates and Lignins, F. Shafizadeh, K. V. Sarkanen and D. A. Tillman (eds.), Academic Press, New York, 1976.
3. R. C. Rothermel, in Thermal Uses and Properties of Carbohydrates and Lignins, F. Shafizadeh, K. V. Sarkanen and D. A. Tillman (eds.), Academic Press, New York, 1976.
4. G. Burton, A. B. Littlewood, and W. A. Wiseman, Gas Chromatogr., Int. Symp. Anal. Instrum, Div. Instrum. Soc. Amer. 6, 193 (1967).
5. F. Shafizadeh, P. S. Chin, and W. F. DeGroot, J. Fire & Flammability/Fire Retardant Chemistry, 2, 195 (1975).
6. F. Shafizadeh and G. D. McGinnis, Carbohyd. Res., 16, 273 (1971).
7. H. M. Spencer and J. L. Justice, J. Am. Chem. Soc., 64, 2311 (1934).
8. J. R. Welker and M. S. Cuvvuri, Report 1932 TR-1, Univ. of Oklahoma (1974).
9. F. Shafizadeh, P.P.S. Chin, and W. F. DeGroot, Forest Sci., 23, 81 (1977).

CONVERSION OF STAGNATED TIMBER STANDS TO PRODUCTIVE SITES AND USE OF NONCOMMERCIAL MATERIAL FOR FUEL

John I. Zerbe

U. S. Forest Products Laboratory
Madison, Wisconsin

I. INTRODUCTION

A perpetual forestry problem is the conversion of unproductive stagnated timber stands to stands of vigorous trees of desirable species. Invariably the major difficulty is paying for the removal of the residual stand so that new growth may be established economically. The prospects for using residual wood material for fuel open up a new source of income to help pay for removal of defective material. Material that was considered noncommercial, now or in the plannable future, may be readily salable for energy values. Examples of types of material that might be economically attractive for fuel are insect- and disease-killed trees, silvicultural thinnings, defective and deformed trees, and trees too small for commercial use on clear-cut areas.

II. LOCATION OF DETERIORATED STANDS

Throughout the country there are large areas of decadent material that, if removed, would have significant forest management benefits. Liquidation of these stands might help to avoid accumulation of unacceptable fuel levels and reduce insect populations. Timber growth could be improved. The health and vigor of timber stands could be enhanced. Provision of access roads could provide for enjoyment of various benefits of timbered areas. Management opportunities for other land uses such as recreation, wildlife habitat, and water could be bettered.

Some areas of the country that are opportune for improved management through removal of defective material are the pine-bark beetle-infected stands of the Rocky Mountains, low-grade stands of hardwoods in the North Central region, spruce budworm-killed material in Maine, hardwoods on pine sites in the South, low-grade hardwood stands in Appalachia, and chapparal in California.

A. Deteriorated Stands in the West

Especially enticing for utilization for energy and other higher value forest products are the Rocky Mountain forests, heavily damaged by insects.

A recent survey [1] estimated that a total of 5.4 billion ft^3 of salvable sawtimber-size material had been killed by insects on lands of all ownerships in the West. Of this amount about 4.6 billion ft^3, or 86%, were on National Forest lands. Particularly promising as an area for demonstrating possibilities for effective use of insect-killed and defective trees is the area surrounding the Targhee National Forest in Southeastern Idaho and Western Wyoming. This ties in to needs for additional generating capacity in Wyoming resulting from the failure of the Grand Teton Dam in 1976. Targhee timber sale slash piles alone could feel enough steam-generating capacity to service the Upper Snake River Valley's present electrical needs. Moreover, both power and power poles could come from the forests.

B. Utilization of Low-Grade Hardwoods

Amother major problem is utilization of low-grade hardwoods. In the South, there are many acres of good land capable of producing rapid growth of Southern pine that are stagnated with accumulations of shrubby hardwoods. These are difficult to harvest and market profitably. In the Appalachian and North Central areas, too, hardwoods are accumulating faster than they are being used, and the growth increments are mainly in lower-grade trees and trees of less desirable species.

III. POTENTIAL USERS OF RESIDUE FUELS

To use this material for fuel, a major question is what is the energy value of material from stagnated timber stands? This is a very complicated question because of the varying situations throughout the country. Because of the large concentrations of the type of material under discussion, the most promising approach to utilization would appear to be installations which would use wood residues in fairly large quantities, possibly from 100 to 1000 dry tons per day. Prime candidates for use of material in these quantities are power-generating plants and process steam producers, or combinations of operations for cogenerating steam and electricity.

A rather extensive study on wood fuel for use by The Green Mountain Power Co., an electric utility in Vermont, was conducted by Battelle Memorial Institute [2]. A preliminary assessment compared the use of wood as a fuel for a commercial electric power plant, with that of clean fossil fuels or fossil fuels with suitable pollution control technology. This study concluded that use of wood for fuel in power generation was feasible and that a demonstration should be conducted.

Previously the Forest Service conducted feasibility studies on utilizing wood residues for fuel for power generation [3] and as boiler fuel in forest products manufacturing plants [1]. Power generation did not seem favorable in the study which was conducted prior to the Arab States oil embargo in 1973, but even then using more wood for fuel in forest products plants seemed promising. This was confirmed and reemphasized in the second study.

In the studies to date, burning of wood for fuel has been investigated on the basis of combustion and environmental control cost comparisons with other fuels. There has been little consideration of more efficient use of fuel by so-called cogeneration or district heating approaches. However, increased efficiency with generation of both electricity and process steam or electricity and domestic heat might be a key to more effective use of wood fuel. A promising possibility would be for generation of power from waste wood by a utility and sale of lower pressure steam exhausted from the turbine to a neighboring industrial user, possibly a paper mill. The cogeneration concept is addressed in detail in a subsequent paper in this volume authored by L. N. Johanson and K. V. Sarkanen.

The second alternative would be the production of lower pressure process steam as a single product.

For any end use, the major problem is removal of material on problem stands, transporting it, and processing it for fuel use economically. Installation of new wood fuel-burning capacity must be justified on the basis of satisfactorily solving this problem, and determining the adequacy of a continuing wood fuel supply over the lifetime of the installed equipment. Timber owners can favorably influence the installation of wood-burning capacity by developing reliable data on availability and anticipated cost of the wood material from stagnated stands over time.

Although overall national studies indicate feasibility for installation of wood-burning facilities to use material from stagnated stands and other residues, actual provision of power generation, steam generation, or cogeneration plants must be justified for specific local situations.

IV. EXAMPLE OF WOOD FUEL USE FOR POWER GENERATION

A recent publication [2] provides a method for analysis of return on investment (ROI) for such installations based on costs for equipment, fuels and power, operation and maintenance, property taxes, and insurance. This analysis procedure might be used to determine the economics of a 50 megawatt (MW) capacity generating plant in an area with deteriorated stands that might be harvested over a long term. The following assumptions might be made:

(1) Cost of fuel delivered to plant - $35.00 per ovendry ton.

(2) Cost of steam-generating capacity - $30.00 per pound per hour of steam.

(3) 1-pound steam per hour - 1000 Btu/hr.

(4) 10-pound steam per hour - 1 kilowatt hour (kwh)

(5) Fuel requirements for 50 MW plant - 250,000 dry tons of wood annually.

(6) Plant should have capacity to burn 50 dry tons of residue per hour (438,000 dry tons annually).

(7) Steam output capacity - 400,000 lb/hr.

(8) Cost of electrical generating capacity - $300/kwh.

(9) Capital investment for burning wood - equivalent to capital investment for burning coal.

(10) Fuel value of wood - 8500 Btu/oven dry pound.

(11) Efficiency of burning wood - equivalent to efficiency for burning coal.

(12) Cost of coal - $2.50/10^6$ Btu.

(13) Annual operating and maintenance costs for wood system - equivalent to annual operating and maintenance cost for coal system.

Table I presents the cost comparison based upon those assumptions. Table II presents the cash flow analysis and Table III presents the analysis of present worth. These tables illustrate the economic problems and opportunities associated with fuel wood utilization for generating electricity

TABLE I. Cost Comparison for Fossil and Residue Fuels

	System using fossil fuel	*Residue fueled system*	*Difference in cost*
Initial investment (purchase and installation)	*$27,000,000*	*$27,000,000*	*0*
Annual fuel costs	*$10,625,000*	*$ 8,750,000*	*-$1,875,000*
Annual operating and maintenance costs	*$ 200,000*	*$ 200,000*	*0*
Annual residue disposal costs	*0*	*0*	*0*
Expected useful life	*10 years*	*10 years*	

TABLE II. Cash Flow Analysis

Year	Costs for residue system	Costs for fossil fuel system	Net change in costs
0	$ 27,000,000	$ 27,000,000	$ 0
1	8,750,000	10,625,000	- 1,875,000
2	9,187,500	11,156,000	- 1,968,500
3	9,646,875	11,714,000	- 2,067,125
4	10,129,000	12,300,000	- 2,171,000
5	10,635,000	12,915,000	- 2,280,000
6	11,167,000	13,560,000	- 2,393,000
7	11,726,000	14,239,000	- 2,513,000
8	12,312,000	14,950,000	- 2,638,000
9	12,927,000	15,698,000	- 2,771,000
10	13,574,000	16,483,000	- 2,909,000

TABLE III. Calculation of Present Worth at 10%

Year	Decrease in cost	-	Compound interest factor	=	Present Worth
1	$ 1,875,000		1.100		$ 1,705,000
2	1,968,500		1.210		1,627,000
3	2,067,125		1.331		1,553,000
4	2,171,000		1.464		1,483,000
5	2,280,000		1.611		1,415,000
6	2,393,000		1.772		1,350,000
7	2,513,000		1.949		1,289,000
8	2,638,000		2.144		1,230,000
9	2,771,000		2.358		1,175,000
10	2,909,000		2.594		1,121,000
					$ 13,948,000

Based on the overall cost of the power plant with the assumptions for the relative costs of the wood and coal-burning systems as made, the ROI would be less than 10%. With all the uncertainties in the assumptions, the low rate of return in this particular example would not justify the risks involved in installing a wood-burning electricity generating system. However, the potential for a particular area with the right combination of lower wood fuel costs and/or higher fossil fuel costs, together with possible cash credits for better land management with the use of wood that is still good, and as the prices for fossil fuels increase, the prospects for using more wood fuels are becoming better.

V. PLANS FOR USING WOOD FUEL FROM STAGNATED STANDS

Currently the U. S. Forest Service is conducting a site-specific study for the U. S. Energy Research and Development Administration (ERDA) on the availability of residues, including deteriorated timber stands, in Northern Wisconsin and the upper peninsula of Michigan. We are also determining the specific present and future requirements of 10 pulp and paper mills in the area for energy. We will develop a computer simulation program to demonstrate how residues can be most economically harvested, processed, and transported to the pulp and paper mills and be used as fuel to serve the energy needs of the mills.

With the availability of a realistic computer simulation program for this area, we will have a good tool to study the major cost-influencing factors in the harvesting and use of forest residues. With a better understanding of the makeup of the costs, we will be better able to concentrate on research to reduce the costs in making residues available. The computer simulation program should also be adaptable to other areas of the country and should be generally useful in making forest residues available more readily and more economically.

On May 20, 1977, ERDA issued a Request for Proposal for the utilization of a minimum of 1000 dry tons of biomass per day for fuel in generating electricity or process steam. I believe waste wood and, in particular, wood from stagnated stands, is logically well suited to this demonstration and others of its type. These demonstrations might well be the first steps in using forest residuals to augment the fuel supply from wood which is now confined in industry to burning sawmilling, pulping, and other plant wastes.

REFERENCES

1. USDA Forest Service, "The Feasibility of Utilizing Forest Residues for Energy and Chemicals." A report to the National Science Foundation and Federal Energy Administration, 1976.
2. E. H. Hall, C. M. Allen, D. A. Ball, J. E. Burch, R. N. Conkle, W. T. Lawhon, T. J. Thomas, and G. R. Smithson, Jr., "Comparison of Fossil and Wood Fuels." Battelle-Columbus Laboratories, Columbua, Ohio.
3. John B. Grantham, Eldon M. Estep, John M. Pierovich, et al, "Energy and Raw Material Potentials of Wood Residues in the Pacific Coast States" A summary of a preliminary feasibility investigation. USDA Forest Service General Technical Report PNW-18, Portland, Oregon.
4. Thomas H. Ellis, "Economic Analysis of Wood- or Bark-Fired Systems." Forest Products Laboratory, Forest Service, U.S. Department of Agriculture, Madison, Wisconsin.

INDUSTRIAL WOOD ENERGY CONVERSION

George D. Voss

American Fyr-Feeder Engineers
Des Plaines, Illinois

I. INTRODUCTION

This paper is based on the experiences of American Fyr Feeder Engineers in the direct combustion and gasification of wood as a source of industrial fuel. The basic process of direct combustion involves the metered feeding of wood into a boiler. There combustion takes place on a refractory grate to produce steam. The process itself is that shown in Fig. 1. The depth of the fuel bed is determined partially by fuel characteristics (sizing, volatile matter content, etc.), but primarily by the moisture content of the fuel. When utilizing a dry fuel, there is a very thin bed with a high degree of suspension burning. When firing a wet fuel, a thicker bed results in the more pronounced layers of chemical reaction shown in the referenced figure. Secondary air is introduced over this fuel bed, and aids in direct combustion of wood and volatiles in suspension. Primary air is fed through the grate and up through the bed

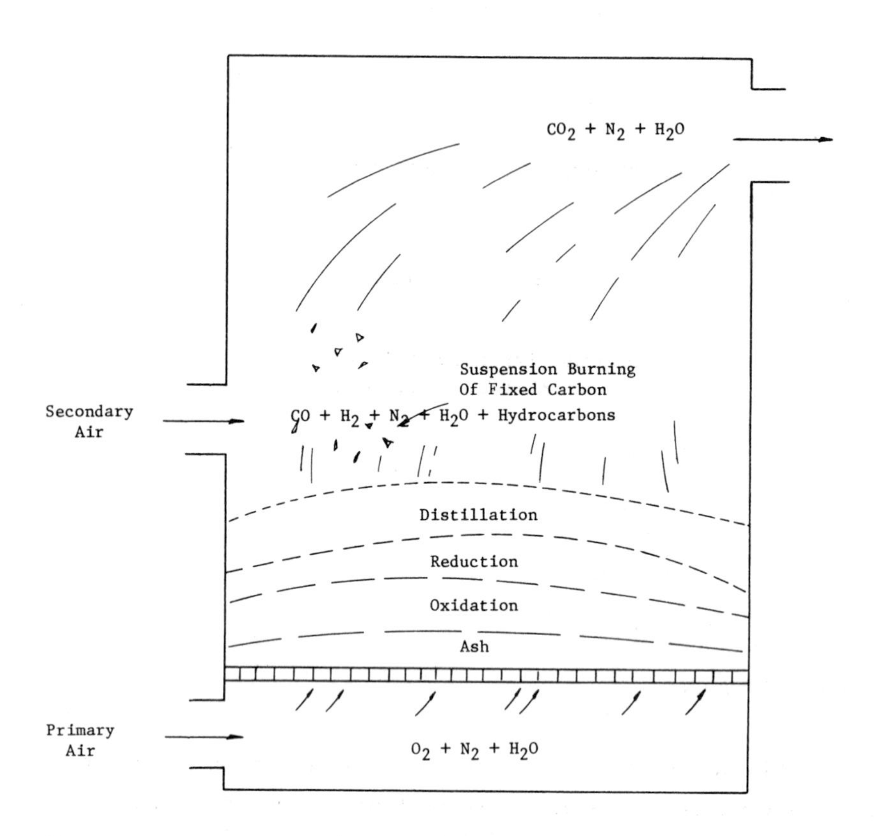

FIGURE 1. Simplified fuel bed firing on stoker-fired boilers.

to aid in the oxidation-reduction reaction and the direct combustion of fixed carbon.

The occurrences of drying in suspension, distillation and burning of the volatile matter, and the burning of fixed carbon are typical of a wood spreader-stoker fired process. There is a very fast response to load changes, and through good combustion controls, the system can be operated very efficiently. Even burning wet fuels of 50% moisture content can be accommodated.

Fixed-bed gasification is an historically proven process [1-4] and is similar to wet-fuel stoker firing as is illustrated in Fig. 2. Our experiences have been with a refractory lined, fixed-bed gasification unit operating under a negative pressure utilizing a jet-pump, gun-type burner arrangement. Drying and distillation occur in the reactor itself, with no burning of fixed carbon. The combustion of gaseous products occurs at the burner located at the boiler.

The primary air needed to sustain the oxidation-reduction reaction is induced through the reactor by the suction of the

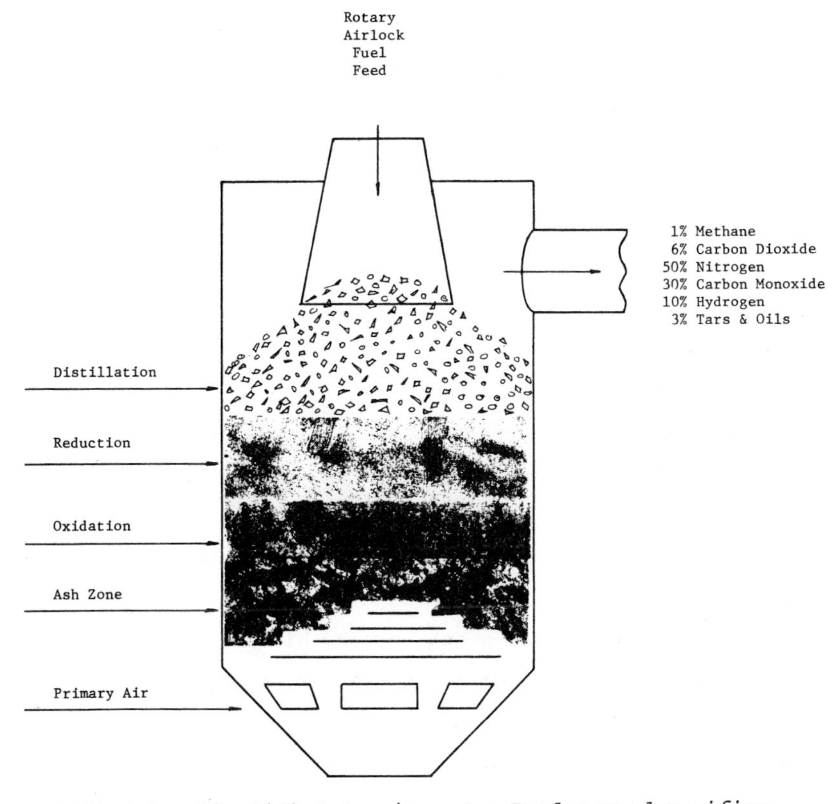

FIGURE 2. Simplified American Fyr-Feeder wood gasification system.

inspirating burner. It could be drawn through the reactor using the boiler induced draft fan. It is important that no combustion air be introduced above the gasifier fuel bed in the reactor, and only introduced at the burner to avoid combustion of the gaseous products consisting of carbon monoxide, hydrogen, and tar and oil vapors. A well designed fixed-bed gasification unit with good combustion controls also can be operated with a good response to load changes and high turndown rates.

II. DIRECT COMBUSTION

Direct combustion as the selection for wood fuel utilization is best justified, from an economic standpoint, in a new boiler installation or in a retrofit situation where the boiler has been designed to fire a solid fuel. A typical arrangement for a complete direct combustion system is that shown in Fig. 3, and will be discussed here in detail. Wood is fed into the receiving hopper through a positive bottom unloader which breaks bridges in the wood fuel silo. It provides a uniform and positive feed to the receiving hopper located at the front end of the metering conveyor. Agitation in this receiving hopper is essential to avoid bridging in the hopper. The level of fuel is controlled by a bin switch which will activate the bottom unloader to fill the hopper.

The fuel is transferred by an auger into the transition piece which feeds it into the injector blower. The fuel is distributed through an air swept housing hitting a deflector casting and spread uniformly over the grate. A high bridge wall at the rear of the grate is essential for maintaining an adequate retention time of the wood fuel in the firebox of the boiler. This practice reduces excessive particulate carryover which would lower furnace temperatures and create a heavy inlet dust loading to the pollution control equipment. After one or several passes

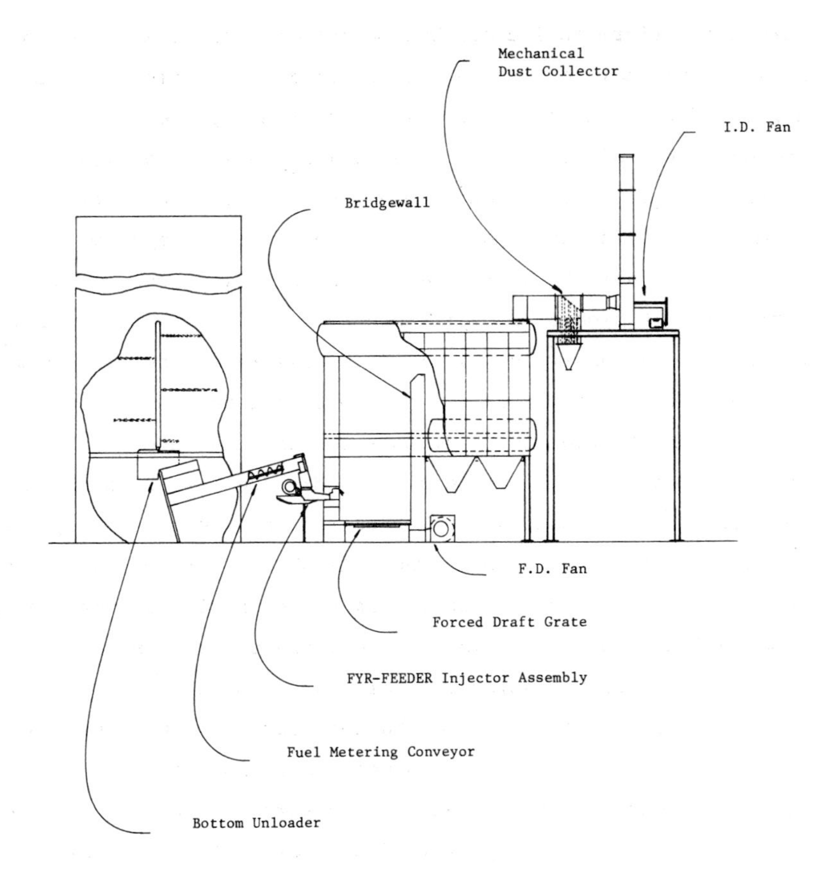

FIGURE 3. A typical arrangement for a complete direct combustion system.

through the boiler tubes, the products of combustion with particulate matter are passed through a multi-tube cyclone for removal of about 90% of the fly ash. Uniform feed of wood into the injector blower is essential to avoid slugging of the material and carryover through the collector and out through the induced draft fan and stack.

Another arrangement of direct combustion is one which involves a blowpipe system feeding the wood fuel into an injector in a similar manner to the previously described system.

Here the uniform fuel distribution into the carburetor before blowing the fuel in through the injector, is critical. A non-uniform fuel feed causes explosions of the dry fuel in the boiler and carryover of black carbon particles. Also in a blowpipe system, there is less control over the amount of air utilized to feed the material into the boiler. The blowpipe system is mainly applicable to a dry fuel which is easily conveyed pneumatically. A high degree of suspension burning occurs, and if the fuel is excessively dry, it is necessary to install steam jets or wet the fuel before introduction into the furnace. Adding moisture to the fuel increases the firebox retention time, and also decreases the tendency of a dry fuel to explode causing the boiler to go positive with a consequent loss of energy. In addition to steam jets to prevent flashbacks through the blowpipe and hopper, there is a blowback damper which will automatically close off the fuel feed line if the furnace goes positive.

Boiler efficiencies are certainly higher utilizing a dry fuel. The energy loss due to the evaporation of moisture is not as significant as when firing a wet fuel. However, there is a minimum moisture content that is necessary, and the addition of water is required to help prevent explosions from occurring, increase firebox retention time, and to minimize particulate carryover. Also, in adding moisture to the fuel or burning a wet fuel, a larger size fuel bed is maintained and gasification of the wood fuel occurs on the grate itself. While the energy needed to evaporate the moisture in a wet fuel is lost, there is some compensation in the oxidation-reduction reaction which occurs on the grate in the same manner as fixed-bed gasification. Efficient boiler plant operations are designed without fuel drying before introduction into the boiler.

The turndown on a direct combustion wood fired boiler is in the range of 4 to 1 or greater depending on the degree of suspension burning and the moisture content of the fuel. To maintain a consistent air-to-fuel ratio and to operate in a balanced draft mode of operation, combustion controls are essential to good boiler plant operation. Combustion controls consist of controlling the amount of fuel and air that is introduced into the boiler in response to steam pressure demands. The balanced draft within the boiler is maintained by operating the induced draft fan damper to maintain a negative draft in the firebox of 0.2 inches.

One problem in wood fired boiler plant combustion control systems is that at low loads there is little control over the air-to-fuel ratio. There are sources of air that are not subject to control. These include the in-leakage from the forced draft fan, the contribution of the injector blower, overfire air header, or ash reinjection system. At low loads the boiler is running at relatively high excess air with no control over the air-to-fuel ratio until one gets into a position where the modulating controls can vary the amount of combustion air introduced into the boiler in accordance with the demand for fuel to meet operating steam pressure.

Another consideration in the operation of the direct combustion unit is maintaining adequate firebox temperatures at low loads to prevent operating off of a hold-fire timer, which, to maintain steam pressure, intermittently turns on and off the fuel feed screw conveyor. The main problem is feeding in fuel to a cool firebox, causing smoke and potential opacity violations as the screw turns on and feeds the fuel into the boiler. One solution to this, especially in the case of a horizontal return tube boiler, is to introduce dilution air in the rear of the boiler to decrease the efficiency of heat transfer through the tubes in the boiler by lowering the flue gas temperatures. This still allows the temperatures in the firebox in front of the boiler bridgewall to be maintained at a higher level. It consumes more wood fuel, which, during the summer months when low loads are present, solves a disposal problem in most wood-related industries.

The ideal solution to increasing fuel consumption during low load conditions is to create an increased steam demand through the application of steam driven equipment. Co-generation is a possibility if there is an adequate plant electrical demand. This concept is discussed in the paper by Dr. Johanson and Dr. Sarkanen. Other equipment like steam-driven air compressors can be installed to take advantage of the excess energy available while maintaining higher firebox temperatures.

III. FIXED-BED GASIFICATION

Fixed-bed gasification of wood fuel offers a tremendous potential application to a variety of industrial energy situations. Gasification offers a certain economic advantage when applied to existing gas and oil-fired boilers. The market is large considering that, for boilers sold between 1963 and 1975 in the industrial boiler size range, 3.9% were designed for coal, 27% for oil, 57% for gas, and 1.4% for wood [5]. Also, gasification offers great potential for the small industrial users in the size ranges below 100 horsepower. In such small cases the total system cost involved in a direct combustion unit, including pollution control equipment, can be high relative to the industry size.

A. System Design

System design, as far as unloading and transfer into the gasifier, is similar to that of direct combustion although the fuel must be fed into the gasifier through an airlock feeder to prevent the introduction of air above the fuel bed. Any introduction of such air would convert the carbon monoxide to carbon dioxide causing a premature release of energy and a possible explosion hazard.

The burning process, as it occurs in the gasifier, involves the ignition of a small amount of fuel at the bottom of the reactor to create the carbonized bed. This must be maintained for continuous gasification. The unit should be completely filled and air tight other than introduction of air through the grate at the bottom supporting the oxidation-reduction reaction. The capacity of the unit is controlled by increasing or decreasing the amount of reaction air introduced through the bottom of the grate. We have experienced turndowns of approximately 6 to 1 on our demonstration units. The reactor is operated continuously but can be shut down almost completely for a period of several hours as long as the carbonized bed is maintained. It will immediately reignite and continually operate once the air is reintroduced through the reactor bed.

Approximately 30% of the total air required for combustion of the gas is introduced through the grate. The remaining 70% is introduced for combustion at the burner. The temperature maintained in the bed of the gasifier is approximately 1650°F which is a cherry red color of wood. The producer gas is measured at a flame temperature in excess of 2500°F. Approximate constitutents of wood gas and natural gas are compared in Table I. It also illustrates the relative combustion properties of wood gas and natural gas.

We have run our demonstration unit burning a variety of biomass commodities including wood chips, pelletized municipal solid waste and corncobs. We have run for periods of 6 to 8 hours with no clinkering effect on the grate.

The volume of the products of combustion of a gasifier is approximately 25% greater than that in direct combustion of oil or natural gas, but we anticipate little boiler derating due to the utilization of a low-Btu gas from wood gasification. A commercial gasifier design

TABLE 1. Combustion Properties of Wood Gas and Natural Gas (Dry Basis)

Property	Wood gas	Natural gas
Flammable limits in air (Volume %)		
Lower	12	4.8
Upper	74	13.5
Composition (Volume %)		
Methane	1	96.0
Ethane	-	3.0
Carbon dioxide	6	0.2
Nitrogen	50	0.8
Carbon monoxide	30	-
Hydrogen	10	-
Tar and oil vapors	3	-
Btu content (Btu/ft^3)	200	1026
Approximate flame temp. °F	3200	3500
Ft^3 dry air/ft^3 gas	1.59	9.65
Btu/ft^3 of gas-air mixture	77.2	96.3
Ft^3 combustible flue gas products per ft^3 gas	1.83	10.6
Btu/ft^3 combustible flue gas products	109	96.7

requires a complete combustion control panel including thermocouples for temperature monitoring and control, and a flame scanner at the burner itself to make sure that ignition is maintained. Several areas in a gasifier design that require additional attention are a rotating grate for continuous ash removal, refractory versus water jacketed vessel construction, and low-Btu gas burner design. The gas is laddened with tars, and the temperature must be kept about 400°F at all times to prevent the tar and oil vapors from condensing out on the equipment. In addition to posing quite a sticky problem with equipment maintenance and upkeep, this condensation causes a loss of energy contained in the tar and oil vapors. This limits the distance which the gas can be piped to the burner at the boiler, and a close proximity of the gasifier to the boiler is recommended. We have utilized a jet-pump, gun-type burner design which uses inspirated air to suck the gas through the gasifier in through the burner with the majority of combustion air introduced at the burner itself.

B. System Applications

Other gasifier applications include the use of multiple gasifier units for larger boilers. They would provide for increased reliability should problems occur with one of the multiple gasification units. Also, production on a smaller scale, as in any industrial process, problems are less severe and a gasifier in several size ranges could be easily manufactured and perfected. Several process developments in gasification are being considered. One is to try a compressor instead of an inspirated type burner to increase the energy in the low-Btu gas, enabling one to pipe it further into the boiler. Another is the design of a device for agitation of the fuel in gasification, to maintain a uniform fuel bed in the distillation layers to avoid bridging in the gasifier itself.

C. Gasification Advantages

Gasification has the advantage of direct energy conversion efficiencies when utilized for application in dryers or industrial furnaces. In addition, the application of the low-Btu gas from wood gasification in engine-generator sets is a process that is being evaluated for somewhat larger scale applications and capital investments. This requires the cooling of the gases, and the subsequent loss of the energy in the tars and oils to prevent the deterioration of the engine internals. The advantage to using the gasifier in a close-couple arrangement is that the process efficiencies are higher. One can maintain the sensible heat of the reaction of the gasifier in addition to the energy available in the tars and oils. The gasifier burns quite cleanly, with the energy from the carbon monoxide and hydrogen more than sufficient to provide ignition of the tars and oils in an after-burner effect at the burner itself.

IV. CROSS COMPARISON

Direct combustion offers a good, reliable method for utilizing wood waste and wood resources as a fuel. Its primary application is in a new installation or a compatible retrofit situation. Gasification offers more potential for a retrofit application on existing oil or gas-fired boilers. It can be utilized in direct application for dryers and industrial furnaces, and, on a larger scale, may be coupled with internal combustion engines or turbines and generators for the direct production of electricity. Although direct combustion is traditional, an efficiently designed system using good combustion controls and fuel metering and feeding systems are more recent developments. They make utilization of wood a very attractive alternative to unavailable or unreliable sources of conventional fossil fuels.

Wood is also a much cleaner source than the alternative solid fuel - coal. The ash content generally averages 1%, and the sulfur content is negligible making the attainment of environmental standards an easier task.

Gasification is an appealing and reborn technology that has potential application to a variety of situations. The most useful and beneficial application for gasification is for the small user who does not have the large energy demand that justifies the larger, though generally affordable, capital investment for a direct combustion unit.

Regardless of which system is utilized, the total system economics are very favorable at current energy prices. They will be even more desirable in the future considering predicted energy costs and trends. A case in point is a recent installation in Minnesota involving a rose greenhouse where fuel is being purchased from a total tree chipper at $12.00 a ton. This firm is installing a complete direct combustion system as shown in Fig. 3. The antitipcated payback time on the entire system will be three years. What is even more important to this customer is the reliability of wood as a fuel. This firm was threatened with the loss of oil supplies last winter, which can be translated into a several hundred thousand dollar loss in greenhouse roses.

In a comparison of the process efficiencies of direct combustion and gasification, one must look at the total system efficiency. Boiler efficiency exclusively is evaluated in the case of direct combustion while the total combined efficiency of the gasifier and the boiler must be considered in gasification. Wood fuel fired boilers can operate at 75% to 80% efficiency, and, with high moisture content fuels, may drop to 65% efficiency. A well designed fixed-bed gasifier can operate at an efficiency of 90% or higher. The losses are minimal, assuming a well designed grate for continuous removal without carbon loss in the ash, slow velocities through the reactor itself to minimize particulate carryover, the avoidance of an excessive amount of fines in the fuel of the reactor, and maintaining adequate gas temperatures to avoid condensation of the tars and oils.

Gasification is most efficient in a close-coupled design where the sensible heat of the reaction and the tars and oils contribute to the Btu content of the gas. Taking an average of 90% gasifier efficiency, and assuming the efficiency of burning this low-Btu gas in the boiler is on the order of 75% to 80%. This would mean the total process efficiency is 67 1/2% to 72%, which is roughly comparable to the efficiencies obtained in direct combustion.

When looking at the efficiency of a system, whether it be direct combustion or gasification, one must consider the application and availability of fuel in a particular situation. In the case in the wood products industry, as well as industries in the vicinity of a wood products industry, there is an abundance of wood waste and wood as a local fuel. It is quite different from the situation of expensive gas or oil fired installations. The critical factor in developing systems for small and intermediate sized industrial users is the capital investment. The return on this investment for any wood energy conversion system may be good, but the magnitude of the capital investment must be kept to a minimum due to the size of the industry itself.

This means optimizing materials transfer, avoiding fuel drying before introduction into the boiler or gasifier, and other means of providing simpler but effective equipment for the conversion of wood to energy. The flame temperature of a gasifier at the burner is 2700°F instead of 3200°F due to the moisture content of the fuel. It does not really matter in quite a number of situations. The introduction of more fuel to make up for the energy loss and the evaporation of the moisture, or the decreased efficiency of the boiler are not as crucial as the capital investment and the additional maintenance and operating expenses associated with complex wood energy conversion systems. In the initial stages of developing a national program for wood energy conversion, one must look at the vast market of small and medium sized users that have limited capital available, and have the most gain from the installation of an affordable wood-fired energy conversion system.

In an environmental comparison of direct combustion versus gasification, there is a decided advantage toward gasification. The gasifier burns more cleanly than a direct combustion unit.

The gasification process involves no suspension burning and the ash is primarily removed through the grate located at the bottom of the reactor vessel. There is some particulate carryover, depending on the velocity through the gasifier and the particle sizing of the wood fuel that is being gasified. However, even this is at a minimum and the effect of the combustion of the carbon-monoxide and hydrogen, at the burner, is to act as an afterburner for the tars and oils. It also aids in the secondary combustion of any carryover particulate matter.

For direct combustion in a spreader-stoker fired boiler, the dust particulate loadings generally require a multiple tube cyclone efficiency of 80% to 90%. The exact efficiency depends on the environmental regulation in that state. Some states, including the State of Illinois, have a regulation of 0.1 pound per million Btu input, which requires a higher efficiency particulate collection device like a scrubber. In most cases, a small to medium sized industry can not afford and can not handle the problems associated with creating a water pollution problem from an air pollution problem. Baghouses have been tried in several locations with rather disappointing results due to the potential for fires and explosions. However, wood is a low ash, low sulfur fuel and attaining most environmental standards in the United States is accomplished with a multiple tube cyclone collector. The real problem with the conversion of wood to energy is the inconsistent development and enforcement of state environmental regulations and the fact that the regulations rarely evaluate environmental impact with any objective measure of comparative severity of impact [6].

V. CONCLUSIONS

A direct combustion boiler system for wood fuel energy conversion is an available and reliable process that can meet industrial plant energy demands. Simplicity and cost effectiveness in design are essential to providing a workable and affordable wood fuel boiler plant operation. Generally, the technology of adapting a wood fuel gasification system to an existing gas or oil fired boiler is relatively simple, and we can see no problems in this type of application.

REFEREMCES

1. N. E. Rambush, Modern Gas Producers, Van Nostrand Company, 1923.
2. H. M. Bunbury, The Destructive Distillation of Wood, Benn Brothers Limited, 1923.
3. J. D. Ross, "Bibliography of Wood Distillation, 1907-1953," Oregon Forest Products Laboratory, April, 1955.
4. Janina Nowakowska and Richard Wiebe, "Bibliography on Construction, Design, Economics, Performance, and Theory of Portable and Small Stationary Gas Producers," Northern Regional Research Laboratory, October, 1945.
5. E. H. Hall, C. M. Allen, D. A. Ball, J. E. Burch, H. N. Conkle, W. T. Lawhon, T. J. Thomas, and G. R. Smithson, Jr., "Comparison of Fossil and Wood Fuels," EPA-600/2-76-056, March, 1976.
6. Committee on Renewable Resources for Industrial Materials, Board on Agriculture and Renewable Resources, Commission on Natural Resources, National Research Council, "Renewable Resources for Industrial Materials," National Academy of Sciences, 1976.

THE PYROLYSIS-GASIFICATION-COMBUSTION PROCESS: ENERGY EFFECTIVENESS USING OXYGEN VS. AIR WITH WOOD-FUELED SYSTEMS

David L. Brink and Jerome F. Thomas

Forest Products Laboratory
University of California
Berkeley, California

George W. Faltico

Kaiser Engineers
Oakland, California

I. INTRODUCTION

Biomass may be viewed as the total living product of photosynthesis. Fossil fuels are the metamorphosed products of biomass. The United States is richly endowed with the resources for producing biomass, and has excelled in learning how to manage this resource. From it we harvest food and fiber which satisfy our needs and aid other nations. Until the early part of this century various components of the biomass, especially wood, have supplied a substantial part of U.S. energy needs. However, with the advent of coal and then cheap petroleum and natural gas, its use declined. Only a small percentage of our national energy needs have been supplied from wood in recent years.

Based upon such compelling considerations as national security, economic health, avoidance of energy emergencies, and prudent use of our remaining fossil fuels; as a nation, we should be totally committed to develop every available alternative energy resource. To do less is to invite crises ranging in scope from regional to international.

Biomass constitutes one alternative energy resource. While food and fiber products produced will continue to be used for their highest values as matertials, the vast amounts of residues, for which there have been few commercial markets, often are wasted. Such residues can be used for or converted into fuels as fast as appropriate facilities can be constructed. Moreover, by establishing new goals and programs, the U.S. can maximize and integrate growth of biomass for production of that fraction of the plant useful only as fuel, as well as for food or fiber. For example, with complete stocking and proper management of our commercial forest lands, it has been estimated [1] that annual growth could be tripled in fifty years. With intensive and enlightened multiple use management of our forest lands, increased production of all products, including those obtained by harvesting timber, will follow. Utilization of biomass as fuel has been accomplished largely through direct combustion. The technology for direct combustion is being improved by increasing thermal efficiency and by reducing environmental impact as is shown in the paper by Mr. Voss. Development of an alternative method of processing, involving pyrolysis and gasification, was essentially suspended after the 1930s and revived on a significant scale only during the last decade.

Fuel efficiency and environmental impact were largely ignored in the development of pyrolysis processes in previous decades. Now in any consideration, environmental compatability of a process will be a necessary criterion of any practical design. Energy efficiency may be ignored again in processes designed to be commercialized in the shortest possible time. However, with the major costs of residues being in their collection, and with the inevitable escalation in value that will occur as demands for such residues increase, energy efficiency will become one of the critical considerations in any method of utilizing this fuel source. Thus energy efficiency is only one of a number of factors that should be taken into account in a complete assessment of a system. Such an analysis should include evaluation of economic feasibility as well as technical feasibility of any new unproved components.

The scope of the evaluation of pyrolysis-gasification made in this paper is confined to the pyrolysis-gasification-combustion (PGC) system previously described [2]. Moreover, it is restricted to a specific set of conditions which, necessarily, have been assumed. The evaluation is carried out by comparison with a system employing direct combustion of the hogged fuel. Even with the limitations imposed, variations are involved which are beyond the scope of this paper. Such factors should be evaluated in any total process assessment. Often in discussions concerning the generation of energy using a pyrolysis-gasification system, the all inclusive a priori judgment is made that such systems have a lower energy efficiency than direct combustion. This question is addressed here. Differences in the equipment used in the systems compared, and the significance of these differences with respect to each process are noted.

II. THE PYROLYSIS-GASIFICATION-COMBUSTION

Pyrolysis-gasification-combustion, as a unit process, has been described previously [3,4,5]. In this system, organic matter is completely converted to gaseous products and an inorganic ash. The gaseous products derived comprise a simple mixture of hydrogen, carbon monoxide, methane, carbon dioxide, and water vapor. Composition of the mixture is largely independent of feed compositions although the latter may vary over a wide range [6]. This is due to the simplicity of the final PG gas mixture and the rapid rate at which the major products equilibrate under PG reactor conditions, principally according to equation (1).

$$CO + H_2O \rightleftharpoons CO_2 + H_2 \quad (1)$$

When sulfur-containing fuels are used, depending upon the inorganic content and composition of the fuel, a portion of the sulfur will appear as hydrogen sulfide [3,7]. No sulfur-containing organic products can survive the conditions used in the PG reactor (second stage). Thus, gas treatment techniques can be employed to remove the low concentration of hydrogen sulfide present in PG gas [7], thereby avoiding the necessity of treating the combustion gas which is greater in amount than the PG gas by a factor of 4. On combustion of the PG gas, trace amounts of remaining hydrogen sulfide are converted to sulfur dioxide. Thus, the so-called total reduced sulfur is eliminated and sulfur dioxide concentrations are reduced to very low amounts in the final combustion gas. Moreover, in the PG gas treating process, particulate material can be removed efficiently. It has been predicted electrostatic precipitators, used in conventional processes for the removal of particulates, will be unnecessary.

III. SYSTEMS EMPLOYING THE PGC PROCESS

Process flows have been calculated for schematic designs of systems using the PGC process for converting organic matter to PG gas. This includes a solids-handling system (e.g., for wood [2] which represents a low-ash content feed containing essentially only C, H, and O) and a liquid-handling system (e.g., kraft black liquor [7] which represents a high-ash content feed containing S and Na in addition to C, H, and O). The system using wood or an equivalent type organic residue is further developed herein. A schematic diagram for this system is shown in Fig. 1. The flow of any gas stream and its enthalpy, as an entensive property, is designated H. The heat loss from each equipment item is designated Q. Total steam production (H_{13}) of the PG boiler is delivered to the throttle of a complete expansion, condensing turbine (not shown) and the condensate from this turbine is returned to the system as boiler feed water (H_{10}).

In this process flow, hogged fuel is introduced to the first stage reactor where moisture is evaporated and the total feed stream is brought to the designated temperature before being discharged to the PG reactor (second stage). Heating is indirect, using the entire exhaust stream from the waste heat boiler. In the second stage the reaction mixture is brought to the specified discharge temperature by admission of the specific amount of oxygen which will effect the temperature rise required by combustion of components in the charge flowing through the reactor. Two systems are presented in this study: one in which air is introduced, and a second in which gaseous oxygen is introduced. Under the conditions attained in the PG reactor, the organic fraction of the fuel is pyrolyzed and gasified to yield the pyrolysis gas. The latter is discharged at the specified temperature to a waste heat boiler in which feed water is preheated. Pyrolysis gas is discharged from the waste heat boiler to the heating jacket of the first stage reactor at a temperature such that the necessary heat transfer can take place to bring the mass flowing in this reactor to the specified temperature.

The gas temperature is brought below the dew point temperature so that condensation of water vapor is initiated in the reaction jacket and is continued as the gas and condensate are discharged to the boiler feed water heat exchanger. In the cooling process, particulate is removed by a nucleation phenomenon. Scrubbing of the gas is carried out if necessary before being

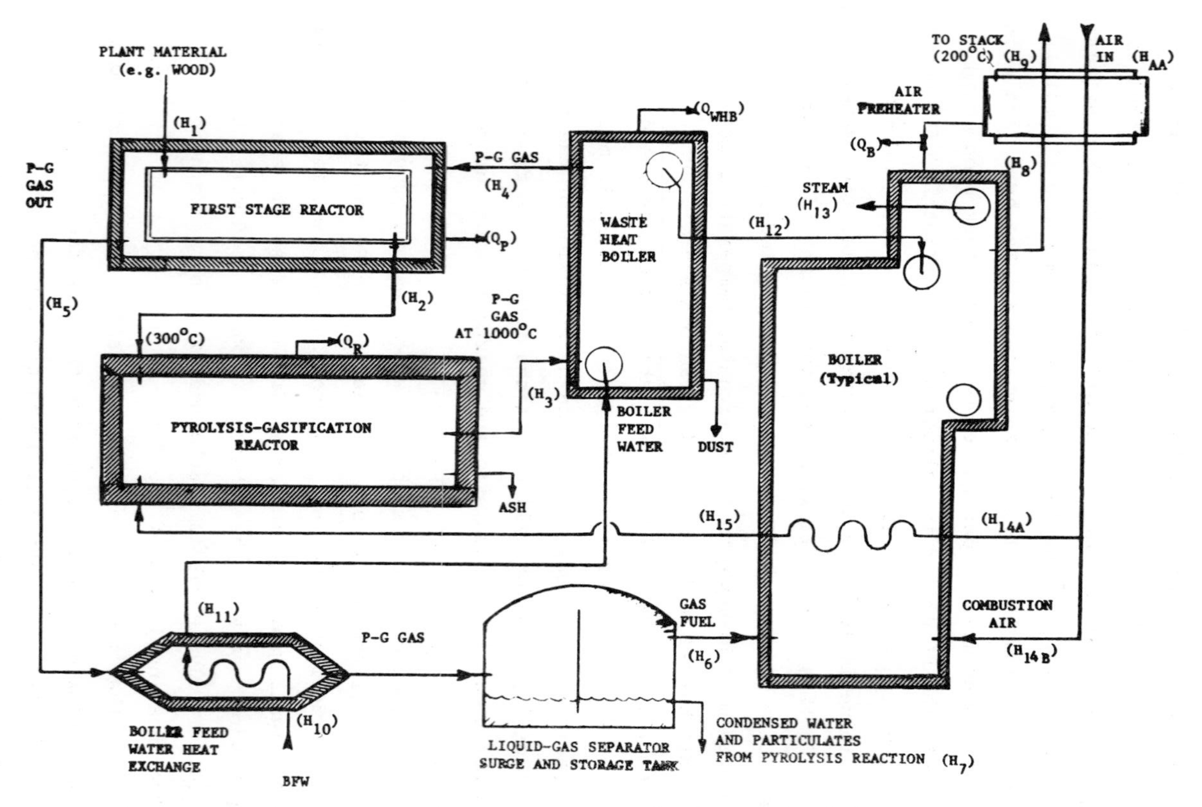

FIGURE 1. *Pyrolysis-gasification-combustion of plant material.*

discharged to the gas storage tank. Boiler feed water is made up of the condensate from the turbine. The gas storage tank furnishes surge capacity for the PG boiler. Condensate discharged from the heat exchanger or the gas storage tank is free of organic contaminants since these do not survive conditions in the PG reactor. The PG gas is introduced to the PG gas boiler on demand with air appropriately preheated by exhaust combustion gases from the boiler. Steam generated is consumed in the turbine-generator as noted above.

Unconventional equipment in this flow are the first stage and PG reactors. A number of devices might be adapted with modification to carry out the functions described. The first stage is a drier and is not treated further here. The high temperature PG reactor may be a shaft, rotary, or fluidized bed type reactor.

A. Mass and Energy Balances

The key to developing the mass and energy balances in this system concerns the conversion of the organic feed into a mixture of simple gases with the accompanying transformation of chemical energy into thermal energy (e.g., the balances around the PG reactor).

1. The PG Reactor

The basis for calculating the mass and energy balances has been presented [8,9] and the method has been applied specifically to wood [2]. In the present paper the method, again applied to wood, is detailed. This study includes the effect of oxygen as well as air in the PG reactor. The more detailed treatment of the boiler-turbine-generator system improves the accuracy of the comparison of the energy efficiencies of the systems considered.

In calculating the mass and energy balance around the PG reactor, it is assumed that wood of a given composition is converted to gas at a given composition. The gas composition is dependent upon the amount of oxygen required to increase the temperature of the feed to the temperature of the products, to supply the heat of reaction, to carry out the indicated conversions and to supply heat lost from the reactor. The composition of materials in percentages of weight fed into the PG reactor are:

Wood - C = 49.25%, H = 6.58%, O = 43.87%, Ash = 0.30%

Air - O_2 = 23.142%, N_2 = 75.519%, Ar = 1.287%, CO_2 = 0.052%

Moisture content of dry air = 0.541% (assumed)

Oxygen = O_2 = 100%

Composition of the PG gas can be determined after making certain assumptions. On gasification of organic matter the amount of methane has been found to be a reasonably consistent value [6]. It has been assumed that this is due to an equilibrium in reaction [2] being approached under conditions used.

$$CH_4 + H_2O \rightleftharpoons CO + 3H_2 \qquad (2)$$

Accordingly, a simplifying assumption has been made that 0.0265 kg of methane are produced per kg of OD wood based on the previous data. Since potassium is a major inorganic cation found in wood ash it has been assumed the ash is potassium carbonate. Then the elements to be equilibrated in a pyrolysis gas, based on wood feed at 110.30% moisture (dry basis), or 52.45% (green basis), are: C = 47.26%, H = 18.26%, and O = 141.83%, on the dry basis, respectively. Determination of the equilibrium constant of reaction (1) at 1273°K is given in equation (3).

$$k = \frac{[H_2][CO_2]}{[CO][H_2O]} = 0.65 \qquad (3)$$

This equilibrium constant has been discussed (2) and is used in this paper.

The theoretical amount of H_2, CO, CO_2, and H_2O derived from the quantities of C, H, and O to be equilibrated without the introduction of air can be obtained by simultaneous solution of equations (3), (4), (5), and (6).

$$\text{kg atoms H} = 0.992162^{*}\ W + 0.111894\ Z \qquad (4)$$

$$\text{kg atoms C} = 0.035702\ X + 0.022722\ Y \qquad (5)$$

$$\text{kg atoms O} = 0.035702\ X + 0.045444\ Y + 0.055508\ Z \qquad (6)$$

where $W = H_2$ (kg), $X = CO$ (kg), $Y = CO_2$ (kg) and $Z = H_2O$ (kg).

a. Air System. The equilibrium mixture of gases obtained with the introduction of a specified amount of air can be calculated by assuming that equal percentages of CO and H_2 are oxidized by introduction of air. The amounts of the air components (O_2, N_2, Ar, CO_2, plus water vapor as indicated) are added to the theoretical amounts of the components calculated as described and the new elemental balance is determined for C, H, and O. The amounts of H_2, CO, CO_2, and H_2O formed at equilibrium are then recalculated by equations (4), (5), (6), and (3) using the new balance. The sensible heat of the equilibrated gases plus N_2, Ar, and constant amounts of CH_4 and ash are calculated and with the heat of combustion of H_2, CO, and CH_4 provide the enthalpy, H_3, of the PG gas flowing from the PG reactor.

Having calculated H_3 under the conditions outlined, it is then possible to complete the energy balance around the PG reactor. Heat loss is calculated from the relationship, $Q_R = 0.01\ (H_3 + Q_R)$. The value of H_2 can be calculated since temperature of this stream is specified at 300°C and, for purposes of this work, the simplifying assumption is made that no thermal degradation of wood has taken place. The temperature of the preheated air introduced into the PG reactor as H_{15} has been specified at 602.6°C. The amount of air for the specific calculation is specified as discussed. Therefore H_{15} can be calculated. The energy balance around the PG reactor, $H_2 + H_{15} = H_3 + Q_R$, will be satisfied when the correct amount of air is used.

* Coefficient used in these calculations was 1.0.

This amount of air is obtained by iteration using the series of calculations outlined.

b. Oxygen System. In calculation of the material and energy balances around the PG reactor using oxygen, the only change made in conditions is the replacement of air with oxygen and the elimination of H_{15}. Oxygen is introduced at the datum temperature.

2. The PGC - Turbine System

The material and energy balances for the overall PGC - turbine system can be calculated after these balances have been determined around the PG reactor. The system is described above, and, excluding the turbine-generator, is shown schematically in Fig. 1.

a. Material Balances. The material balances for both the air and oxygen systems are given in Table I. Additional assumptions are necessary in calculating the material balances. In the combustion of the clean PG gas it is assumed that 7.5% of excess air, normally required in equivalent gas fired boilers, is used. Also, the amounts of condensates produced on cooling the PG gas in the system are based upon the temperature of the PG gas and condensate leaving the particular equipment. These temperatures, indicated in Table I, are calculated after the enthalpy of the stream leaving the unit has been determined in the energy balance calculations.

b. Energy Balance. The PGC process energy balances were calculated following the steps outlined in Table II for both the air and oxygen systems. Assumptions made common to both systems are:

Datum temperature = ambient air temperature = 25°C.

Temperature specified for streams:

H_2 = 300°C

H_3 = 1000°C

$H_6 = H_7 = 44°C$
$H_9 = 200°C$
H_{10} (or H_{BFW}) = 77°F (26.1°C), specified as a design parameter of the turbine; i.e., condenser fluid temperature, t_{fx}
H_{13} (steam) = 925°F (496.1°C) at 1300 psig
$H_{14_{A+B}} = 350°C$
$H_{15} = 602.6°C$

The temperature increase of the boiler feed water was specified at approximately 46°C for an assumed flow. Then the enthalpy required, H_5, in order to satisfy the enthalpies of the streams flowing from the boiler feed water heat exchanger, $(H_6 + H_7 + Q_{PPGST}) + H_{11} + Q_{BFWHE} - H_{10}$, was calculated by an iterative procedure. With the values of H_5 set, as given in Table II, H_{11} could be calculated for an assumed boiler feed water flow. Q_P, H_4, H_{12}, and H_{13} were then calculated.

When the energy inputs and outputs of the PG boiler are balanced, the quantity of boiler feed water has been correctly determined. This balance was obtained by an iterative calculation changing the quantity of boiler feed water, with the results obtained as given in Table II. Temperatures established based on the enthalpies of the material flows in the two systems are:

Flow	Pressure psig	Air system °C	Air system °F	Oxygen system °C	Oxygen system °F
H_4	14.696	426.0	798.8	524.7	976.5
H_5	14.696	52.7	126.9	57.1	134.8
H_8	14.696	426.0	798.8	409.5	769.1
H_{11}	1300.0	44.4	112.0	44.5	112.1
H_{12}	1300.0	230.9	447.6	191.2	347.3

Table I. Material Balance Pyrolysis-Gasification-Combustion Using Air or Oxygen

Item	Air System			Oxygen System		
	Flow, kg/MT, (basis = 1000 kg = 1 metric ton (MT) of oven dried wood)					
	In	Out	Product	In	Out	Product
First Stage Reactor						
Wood, Organic	997.0*	997.0		997.0*	997.0	
Ash	3.0*	3.0		3.0*	3.0	
Water	1,103.0*	1,103.0		1,103.0*	1,103.0	
Total	2,103.0	2,103.0		2,103.0	2,103.0	
Pyrolysis-Gasification-Reactor						
Wood, Solids	1,000.0			1,000.0		
Water Vapor	1,103.0			1,103.0		
	2,103.0			2,103.0		
Air						
Oxygen	482.8*	---		460.7*	---	
Nitrogen	1,575.5*	1,575.5		---	---	
Argon	26.9*	26.9		---	---	
Carbon dioxide	1.1*	938.0		---	914.7	
Hydrogen	---	65.3		---	67.1	
Carbon monoxide	---	505.8		---	520.0	
Methane	---	26.5		---	26.5	
Total, dry	2,086.3	3,138.0		2,563.7	1,528.3	
Water vapor	11.3*	1,059.6		---	1,032.5	
Total, wet	2,097.6	4,197.6			2,560.8	
Ash		3.0			3.0	
Total	4,200.6	4,200.6		2,563.7	2,563.8	
Waste Heat Boiler						
PG Gas						
Dry	3,138.0	3,138.0		1,528.3	1,528.3	
Water vapor	1,059.6	1,059.6		1,032.5	1,032.5	
Ash	3.0		3.0+	3.0		3.0+
Total	4,200.6	4,197.6	3.0	2,563.8	2,560.8	3.0
First Stage Reactor Jacket						
PG Gas						
Dry	3,138.0	3,138.0[1]		1,528.3	1,528.3[2]	
Water vapor	1,059.6	381.3[1]		1,032.5	278.2[2]	
Condensate		678.3[1]			754.3[2]	
Total	4,197.6	4,197.6		2,560.8	2,560.8	

Boiler Feed Water Heat Exchanger											
PG Gas											
Dry	3,138.0		3,138.0[3/]			1,528.3		1,528.3[3/]			
Water vapor	381.3		230.6[3/]			278.2		131.5[3/]			
Condensate	678.3		829.0[3/]			754.3		901.0[3/]			
Total		4,197.6		4,197.6			2,560.8		2,560.8		
PG Gas Storage Tank											
PG Gas											
Dry	3,138.0		3,138.0			1,528.3		1,528.3			
Water vapor	230.6		230.6			131.5		131.5			
Condensate	829.0				829.0+	901.0				901.0+	
Total		4,197.6		3,368.6	829.0		2,560.8		1,659.8	901.0	
PG Gas Boiler											
PG Gas											
Dry	3,138.0					1,528.3					
Water vapor	230.6					131.5					
Total		3,368.6					1,659.8				
	Air (7.5% Excess)				Combustion Gas	Air (7.5% Excess)				Combustion Gas	
Oxygen	981.4*				68.5+	1,005.1*				70.1+	
Nitrogen	3,202.7*				4,778.2+	3,280.2*				3,280.2+	
Argon	54.6*				81.5+	55.9*				55.9+	
Carbon dioxide	2.2*				1,807.6+	2.3*				1,806.7+	
Dry		4,240.9			6,735.8		4,343.5				5,212.9
Water vapor	22.9*				896.5+	23.5*				813.9+	
Total, moist		4,263.8			7,632.3		4,367.0				6,026.8
Total, overall process	(Σ*) 8,464.4				(Σ+) 8,464.3	(Σ*) 6,930.7				(Σ+) 6,930.8	

[1/] at 52.8°C [2/] at 57.1°C [3/] at 44.0°C

Table II. Pyrolysis-Gasification-Combustion Process Energy Balance

Energy Flow	M kg-cal/OD Metric Ton of Wood, Combustion in PGC Reactor: Air System	Oxygen System
H_1 (Wood, at datum temperature, 25°C)		
H_c = 1000 kg,[1] wood x 4,730 $\frac{\text{kg-cal}}{\text{kg}}$ =	[4,730.0][2]	[4,730.0]
Sensible heat =	0.0	0.0
H_2 (Preheated Wood + Water Vapor)		
$H_{c\ wood}$ =	[4,730.0]	[4,730.0]
$m_{wood} \int_{25°C}^{300°C} cp\ dt$ = 1,000 x 0.56[3] x 275 =	154.00	154.0
m_{H_2O} ($h^{4/}_{g,300°C} - h_{f,25°C}$) =	782.27	782.27
	5,666.27*	5,666.27*
H_3 (Pyrolysis-Gasification Gas)		
$H_c(H_2 + CO + CH_4)$ =	[3,795.11]	[3,889.48]
Sensible heat, at 1000°C		
$mi \int_{25°C}^{1000°C} cp\ dt$,[3] + m_{H_2O} ($h_{g,1000°C} - h_{f,25°C}$)	2,151.43	1,758.96
i components are: H_2, CO, CO_2, CH_4, and Ash, plus N_2 and Ar in Air System =		
	5,946.54*	5,648.44*
QR = 0.01 x (H_3 + Q_R) =	21.73	17.77
H_{15} (Air to PG Reactor, finally preheated to 602.6°C)		
$mi \int_{25°C}^{602.6°C} cp\ dt$ = 144.35 $\frac{\text{kg cal}}{\text{kg moist air}}$ =	302.77	0.00

i^{th} component	Weight percent
O_2	23.142
N_2	75.519
Ar	1.287
CO_2	0.052
dry air	100.000
water vapor	0.541 (assumed)

Table II. (continued) Pyrolysis-Gasification-Combustion Process Energy Balance

Energy Flow	M kg-cal/OD Metric Ton of Wood Combustion in PGC Reactor Air System	 Oxygen System
H_6 (PG gas to PG Boiler, assume 44°C)		
H_c =	[3,795.11]	[3,889.48]
$m_i \int_{25°C}^{44°C} cp\ dt + m_{H_2O} (h_{g,44°C} - h_{f,\ 25°C})$ =	154.29	88.45
$m_{H_2O} = \frac{MW_{H_2O}}{MW\ dry\ PG\ Gas} \left(\frac{p_{H_2O}}{1 - p_{H_2O}} \times \frac{kg\ dry\ PG\ gas}{MT\ wood}\right)$		
p = partial pressure of water, in atmospheres, in water saturated PG gas at T°C.		
H_7 (Condensate)		
$m_{H_2O} (h_{f,44°C} - h_{f,25°C})$ =	15.63	16.98
$m_{H_2O} = m_{H_2O}$ PG Gas $- m_{H_2O}$ in H_6 stream		
Q_{PGGST} = 0.01 (sensible heat in $H_6 + H_7 + Q_{PGGST}$ =	1.72	1.06
H_9 (PG Gas combustion gases, 7.5% excess air, exit stack at 200°C)		
$m_i \int_{25°C}^{200°C} cp\ dt$ =	281.97	216.98
i^{th} components are CO_2, N_2, Ar, O_2, H_2O		
$+ m_{H_2O} (h_{g,200°C} - h_{f,25°C})$ =	594.11 876.08	539.35 756.33
H_{14A+B} (Air Preheated to PG Boiler, initial to PG Reactor)		
$m_i \int_{25°C}^{350°C} cpdt + m_{H_2O} (h_{g,350°C} - h_{g,25°C})$ =	468.26	346.29
i^{th} components are: N_2, AR, CO_2, O_2		
H_{AA} =	0.00	0.00
Q_{AP} = $(H_9 + H_{14A+B} + Q_{AP})$ =	13.58	11.14
H_8 = (PG Boiler Combustion Gas to AP)		
$= H_9 + H_{14A+B} + Q_{AP} - H_{AA}$ =	1,357.92	1,113.76

Table II. (continued) Pyrolysis-Gasification-Combustion Process Energy Balance

Energy Flow	M kg-cal/OD Metric Ton of Wood	
	Combustion in PGC Reactor	
	Air System	Oxygen System
H_{10} (Boiler Feed Water (BFW)); assumed at 26.1°C =	5.28	5.47
m_{BFW} (by iteration) Air System, 4754 kg; O_2 System, 4928 kg		
H_{11} =	[3,795.11]	[3,889.48]
H_c		
sensible heat (by iteration), T_{BFW} leaving BFWHE at 44.5°C		
H_{11} = sensible heats of $(H_5 - H_6) + H_{BFW} - Q_{BFWHE} - H_7 - Q_{PGGST}$ =	101.30	105.17
Then		
Q_{BFWHE} = 0.01 (sensible heat of $H_6 + H_7 + H_{11} + Q_{PGGST} + Q_{BFWHE}$) =	2.76	2.14
H_5		
H_c =	[3,795.11]	[3,889.48]
+ sensible heats of $(H_{11} + H_6) + H_7 + Q_{PGGST} + Q_{BFWHE} - H_{10}$ =	270.41	208.33
Q_p = 0.01 [sensible heats of $(H_2 + H_5) + Q_p$] =	12.19	11.56
H_4		
H_c =	[3,795.11]	[3,889.48]
+ sensible heats of $(H_2 + H_5 - H_1) + Q_p$ =	1,218.87	1,156.17
Q_{WHB} = 0.01 (sensible heats of $H_3 + H_{11}$) =	22.53	18.64
H_{12}		
H_c =	[3,795.11]	[3,889.48]
+ sensible heats of $(H_3 + H_{11} - H_4) - Q_{WHB}$ =	1,011.33	689.32
Q_{PGB} = 0.01 $(H_6 + H_{12} + H_{14A+B})$ =	54.29	50.13
H_{13} (Steam from PG Boiler) assume 1300 psig; 925°F;		
= $H_6 + H_{12} + H_{14A+B} - H_8 - H_{15} - Q_{PGB}$ =	3,714.01	3,849.65

1/ See Table 1 for mass (mi)
2/ Heats of combustion are given in [ ], Heats of combustion plus sensible heats are denoted by an asterisk (*).
3/ from Ref [11], Table 3-136
4/ h_f, h_g = specific enthalpy of H_2O (liquid) and H_2O (gas) respectively.
5/ $\int_{T_0}^{T_1} cp\ dt$ from equations, Ref [11], Table 3-174.

IV. THE DIRECT COMBUSTION BOILER

The material balance for the direct combustion boiler, based on the same parameters used in the two PGC systems when applicable, is given in Table III. The amount of excess air used, 37.5%, is the average given [10] for a hogged fuel boiler. It is assumed that there would be a small amount of carbonaceous ash as fly ash or in the ash from the furnace grates. It is further assumed that, on the scale being used (e.g., generation of 30,000 kw) both cyclones and an electrostatic precipitator would be required in combustion gas treatment to reduce particulate material to allowable levels. The system, as described for this comparison is shown in Fig. 2. The same boiler system schematic was used previously [2]. In this paper, however, two systems are considered based upon the turbine-generators specified (not shown in Fig. 2). The energy balance for the direct combustion boiler, Table IV, is given in greater detail than previously. The major difference is in the specification of turbine characteristics and operating conditions so that a more accurate comparison can be made with the two PGC-turbine systems. The systems presented in Table IV are for a nonextraction (nx) (e.g., complete expansion) condensing turbine and for a regenerative (x) condensing turbine with four extraction points for boiler feed water heating.

The nonextraction condensing turbine is the same turbine specified for the two PGC systems. Condensate from the turbine condenser is returned to the boiler ($H_{25(nx)}$) at 26.1°C. Since the sensible heat in this stream is minor there is an appreciable loss in steam production in this system.

The regenerative turbine produces a greater flow of steam at the turbine throttle and requires a boiler having a higher capacity for steam generation, but it provides a more efficient system. Boiler feed water is heated to an optimum temperature in this system.

Table III. Direct Combustion Material Balance[1]

	Kg					
	C	H	O	N_2 + Ar	Ash	Total
Input						
Wood*	492.5	65.8	438.7	--	3.0	1,000.0
Water*	--	123.4	979.6	--	--	1,103.0
Air[2]						
O_2	--	--	1,901.9	--	--	1,901.9
N_2	--	--	--	6,206.4	--	6,206.4
Ar	--	--	--	105.8	--	105.8
CO_2	1.2	--	3.1	--	--	4.3
dry air	1.2	--	1,905.0	6,312.2	--	8,218.4
H_2O	--	5.0	39.5	--	--	44.5
Total air*	1.2	5.0	1,944.5	6,312.2		8,262.9
Total Σ*	493.7	194.2	3,362.8	6,312.2	3.0	10,365.9
Output						
Combustion Products						
O_2	--	--	518.7	--	--	518.7
N_2	--	--	--	6,206.4	--	6,206.4
Ar	--	--	--	105.8	--	105.8
CO_2	488.7	--	1,302.1	--	--	1,790.8
dry gas	488.7	--	1,820.8	6,312.2	--	8,621.7
H_2O	--	194.2	1,541.2	--	--	1,735.4
Total gas	488.7	194.2	3,362.0	6,312.2	--	10,357.1
Ash	5.0	--	3,362.8	6,312.2	3.0	10,365.9
Total	493.7	194.2	3,362.8	6,312.2	3.0	10,365.9

[1] Basis: 1000 kg = 1 metric ton (mt) of wood, OD.

[2] 37.5% excess air

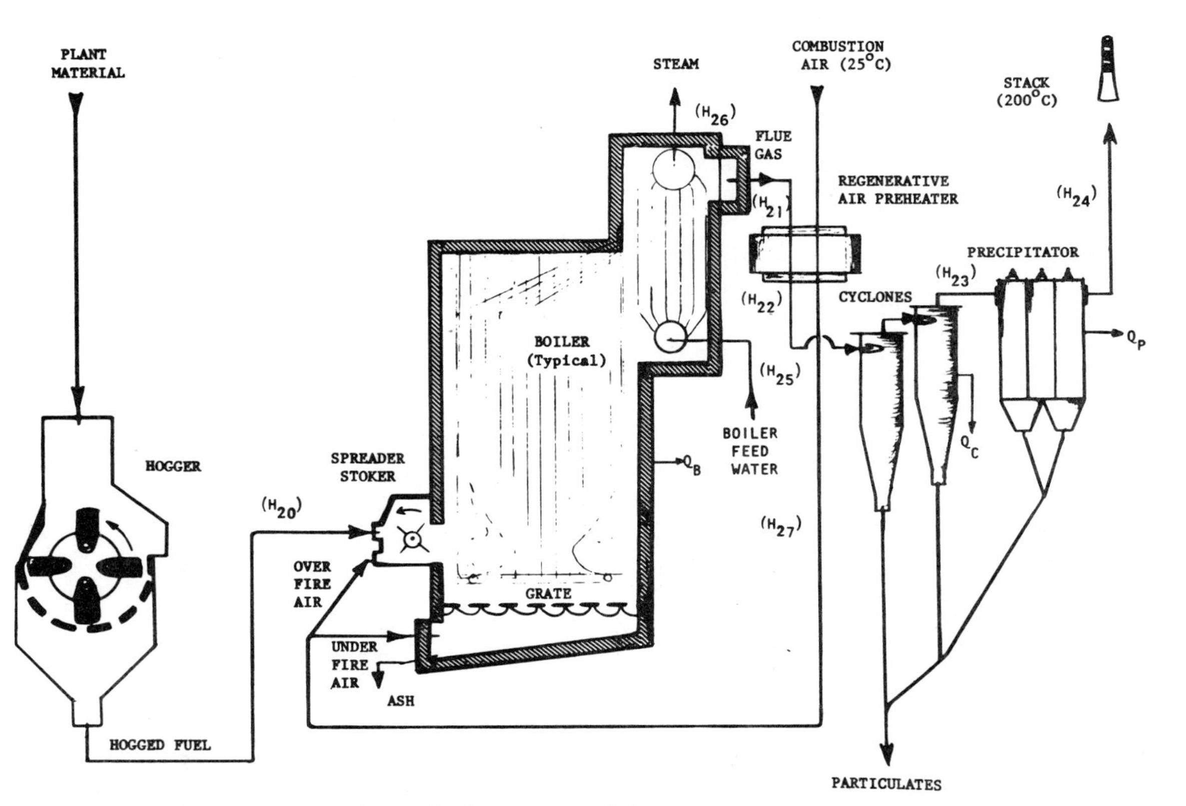

FIGURE 2. Direct combustion of plant material.

Table IV. Direct Combustion Boiler[1/]

Energy Flow		M kg-cal/OD Metric Ton of Wood	
Overall Energy Balance			
$H_{20} + H_{25} + H_{AA} = H_{24} + H_{26} + H_{Ash} +$			
$Q_B + Q_{AP} + Q_C + Q_P$			
H_{20} (wood at datum temperature, 25°C)			
(see H_1, Table I) H_c	=		4,730.00
Sensible heat	=		0.00
H_{27} (Sensible heat of air preheated to 350°C)			
(see H_{15}, Table I)			
$m_i \int_{25°C}^{350°C} cpdt$	=	648.53	
m_{H_2O} ($h_{g,350°C}$[3/] $- h_{g,25°C}$)	=	6.70	
			655.23
H_{AA} (ambient air at datum temperature = 25°C)			0.00
H_{24} (combustion gases, 37.5% excess air, exit stack at 200°C)			
(see H_9, Table I)			
Sensible heat dry gas	=	360.62	
Sensible heat water vapor	=	1,150.00	
			1,510.62
Q_p (heat loss from electrostatic precipitator)			
$Q_p = 0.01 (H_{24} + Q_p)$	=		15.26
$H_{23} = H_{24} + Q_p$	=		1,525.88
Q_c (heat loss from cyclones)			
$Q_c = 0.01 (1525.88 + Q_c)$	=		15.41

Table IV. (continued) Direct Combustion Boiler

Energy Flow		M kg-cal/OD Metric Ton of Wood
$H_{22} = H_{23} + Q_c$	=	1,541.29
Q_{AP} (heat loss from air preheater)		
$Q_{AP} = .01\ (H_{22} + H_{27} + Q_{AP})$	=	22.19
$H_{21} = H_{22} + H_{27} + Q_{AP} - H_{AA}$	=	2,218.71
H_{ash} (neglect sensible heat)		
$m_{carbon} \times h_{c,carbon}$ = 5.0 kg x 8,036 kg-cal	=	40.18
$H_{25(nx)}$ (boiler feed water from turbine condenser)		
$H_{25(nx)} = m_{BFW(nx)}\ (h_{fx} - h_{f,25°C})$		
= 3939 kg (47.09 - 45.09) Btu/lb 555556	=	4.38
$Q_{DCB(nx)} = 0.01\ (H_{20} + H_{27} + H_{25(nx)})$	=	53.90
H_{26} (steam $flow_{nx}$)		
$H_{26(nx)} = H_{20} + H_{27} + H_{25(nx)} - H_{21} - H_{ash} - Q_{DCB(nx)}$	=	3,076.82
$H_{25(x)}$ (boiler feed water from 4th heater regenerative turbine		
$H_{25(x)} = m_{BFW(x)}\ (h_8 - h_{f,25°C})$		
= 5218 kg (394.8- 45.09) Btu/lb x .555556	=	1,013.77
$Q_{DCB(x)} = 0.01\ (H_{20} + H_{27} + H_{25(x)})$	=	63.99
H_{26} (steam $flow_x$)		
$H_{26(x)} = H_{20} + H_{27} + H_{25(x)} - H_{21} - H_{ash} - Q_{DCB(x)}$	=	4,076.12

1/ BFW from nonextractive condensing turbine

2/ See Table III for mass of ith components of input materials and products

3/ h_g, h_f, h_{fs}; specific enthalpies of gas, liquid and turbine condenser liquid, respectively.

V. COMPARISON OF THE SYSTEMS

Based upon the data given in Tables II and IV a direct comparison can be made of the relative efficiencies of the four systems considered. It has been assumed that heat losses can be represented by the calculations given. Other losses in the system are considered to be essentially the same and may be omitted from these calculations. The data presented in Table V comparing the power generated by the four systems is based on the data calculated as described.

Steam generated by the three systems using the nonextraction turbine is directly comparable. It represents the net steam generated by the boiler alone except for the small contribution due to H_{10} and $H_{25(nx)}$, respectively. The direct combustion boiler system provides a substantially reduced flow of steam, as well as power, based upon a given quantity of wood. This has also been expressed as the wood rate per hour required in each system to produce 30,000 kw. Process efficiencies are given as percentages of the net steam produced based on the heating value of the fuel, and of the power generated, based on that theoretically available.

When power generation is the objective, the system using the regenerative turbine normally would be used. This turbine is more expensive and requires a boiler having a higher steam capacity. However, it provides a substantially higher power generation efficiency (24.3%) than a system using a nonextraction turbine (21.3% power generation efficiency). As shown in Table V, the fuel requirement is higher, however, and the power generated is lower for the regenerative system than for either of the PGC systems. The PGC system using air has a power generation efficiency of 25.7% and the PGC oxygen based system has a power generation efficiency of 26.6%. Since the boiler feed water temperature for the PGC system using oxygen is substantially below the optimum, the efficiency of this system could be further improved by using a regenerative turbine. The gain in efficiency using this system has not been calculated.

In the final analysis economics will determine whether the PGC system is viable. A number of critical questions are not answered by the comparison made. These include the designs of the first stage and PG reactors and the boiler feed water heat exchanger. The feasibility of utilizing the heat present in H_4 and H_5, to the extent of the calculations given in Table II, will be dependent upon heat transfer and equipment design. Consideration of these questions is beyond the scope of this paper. The effect of lowering the moisture content of the fuel by predrying would substantially decrease the amount of air and oxygen required in the PG reactor. This would improve the efficiencies of all systems significantly but, in particular, would affect the economics of the oxygen system. Significant reductions in capital costs resulting from the use of a gas fired rather than solid feed fired boiler, the elimination of the electrostatic precipitator, and the capability of using a nonextraction turbine and yet providing boiler feed water at a high temperature are attributes of the PGC systems. These attributes are offset to an extent not determined by costs of the PG reactor, the first stage reactor, the waste heat boiler, the boiler feed water heat exchanger and the PG gas storage tank. A possible advantage of the PGC systems is afforded in the capability of storing the PG gas which can then be used in periods of time requiring peak power generation. Using the oxygen system, in particular, a fuel gas having a medium heating value is produced. It could be transported by pipeline for modest distances.

Table V. Comparative Power Generated

System				
Steam generation	Pyrolysis-Gasification-Combustion		Direct Combustion	
	Air	Oxygen-Air[1/]	Air	Air
Power generation	Condensing-type Turbine Generators[2/]			
	Complete expansion			Regenerative
Steam Flow[3/]				
W_1 (Boiler to turbine throttle)				
kg/mt, OD wood	4,754	4,928	3,939	5,218
lb/short-ton (st), OD wood	9,508	9,856	7,889	10,436
Heat				
M kg-cal/mt, OD wood	3,714	3,850	3,077	4,076
M Btu/mt, OD wood	14,738	15,278	12,210	16,175
M Btu/st, OD wood	13,370	13,860	11,077	14,674
Boiler Feed Water, from	WHB	WHB	turbine condenser	No. 4 heater
Designation on Figures 1 & 2, resp.	H-12	H-12	H-25	H-25
Temperature, °C	249.3	191.2	26.1	215.6
°F	480.7	347.3	79.0	420.0
enthalpy, h_{BFW}	to PGB	to PGB	h_{fx}	h_i
kg-cal/kg	258.64	178.27	26.10	219.3
Btu/lb	465.56	320.88	47.09	394.8
Sensible Heat (H_{12}) M kg-cal/mt	1,011.3	689.3	--	--
SR		Nonextraction[4/]		Extraction[5/]
kg/kwhr		3.37		3.95
lb/kwhr		7.42		8.70
HR		Nonextraction[6/]		Extraction[7/]
kg-cal/kwhr		2,622		2,290
Btu/kwhr		10,418		9,086
PG		Nonextraction[8/]		Extraction[9/]
kwhr/mt, OD wood	1,413	1,464	1,172	1,337
kwhr/st, OD wood	1,282	1,328	1,063	1,213
Wood Rate/hr at 30,000 kw				
mt	23.23	20.49	25.60	22.44
st	23.40	22.59	28.22	24.73
Process efficiency, %				
Basis				
Net steam from fuel[10/]	78.4	81.3	65.1	64.7
Power generated[11/]	25.7	26.6	21.3	24.3

Table V. (continued)

1/ Oxygen used in PGC Reactor; Air used in PG Gas Boiler

2/ Calculations by procedure Ref [10], pp. 4-02 to 4-28; 8-56 to 8-88 and 19-25 to 19-27 made for nominal 30,000 kw rated, condensing type turbines with hydrogen cooling, 0.8 power factor, 3600 rpm; assuming overall engine efficiencies of 76.2% (including 4% exhaust loss, 1.25% mechanical loss, 98.4% generator efficiency) corrected to 76.7% by a factor of .99 for steam superheat. Complete expansion (i.e., no steam extraction for BFW heating) assumed for PGC and DCB. These results are compared with a DCB system using a regenerative-type turbine assuming 4 points of steam extraction for BFW heating other conditions as specified.

3/ Steam Conditions

Throttle - 1300 psig, 925°F (496.1°C), h = 1,451.2 Btu/lb

s = 1,5860, superheat = 346°F,

Exhaust - 1 in. H_g, abs., h_2 = 851.5 Btu/lb

s = 1,5860,

available energy, $h_1 - h_2$ = 599.7 Btu/lb

h_{fx} = condenser liquid enthalpy = 47.09 Btu/lb

h_8 = boiler feed water enthalpy, from final (i.e., 4th) heater

4/ SR_{nx} = nonextraction steam rate = $\frac{\text{3413 Btu/kwhr}}{\text{available energy x overall engine efficiency, corrected}}$

5/ $SR_x = SR_{nx} \frac{(1 + \text{\% increase in turbine throttle steam flow})}{100}$

6/ HR_{nx} = Nonextraction heat rate = $(h_1 - h_{fx})\ SR_{nx}$

7/ HR_x = Extraction heat rate = $[HR_{nx}\ (1 - \frac{\text{\% Reduction}}{100})]$x

[1 - % decrease in exhaust loss];

% reduction of HR_{nx}, 11.5% due to use of 4 extraction points under steam conditions specified and 0.5% increase due to cycle losses; net reduction 11%.
Assume 2% decrease in exhaust losses due to steam extraction.

8/ PG_{nx} = Power generated, nonextraction = $\frac{W_1\ (h_1 - h_{fx})}{HR_{nx}}$

9/ PG_x = Power generated, extraction = $\frac{W_1\ (h_1 - h_8)}{HR_x}$

10/ $\frac{W_1 \times \Delta h}{h_{c,OD\ wood}} \times 100$ $\Delta h = h_1 - h_{fx}$, nonextraction
$= h_1 - h_8$, extraction

11/ $\frac{\text{kwhr/st, OD wood}}{\frac{H_{c,\ OD\ wood}}{3413}} \times 100$ = .020043 kwhr/st, OD wood

VI. CONCLUSIONS

It has been demonstrated that a PGC system, designed to utilize sensible heat effectively, may provide higher efficiencies with respect to both steam generation and power generation than a system employing a hogged fuel boiler. Capital costs of equipment have not been estimated. They remain an important question in determining feasibility of the PGC process. Significant questions are unanswered concerning the design of several equipment items in the PGC system. Further studies to establish optimum parameter settings for the system, particularly the water content of the feed, should be performed. This study suggests that a PGC system may offer benefits that cannot be obtained with a direct combustion system. In view of the serious problems concerning energy supply, systems of the kind envisaged could make a significant contribution in efficiently utilizing one alternative energy resource, plant residue.

REFERENCES

1. Renewable Resources for Industrial Materials, A Report of the Committee in Renewable Resources for Industrial Materials (CORRIM), Board on Agricultural and Renewable Resources, National Academy of Sciences, Washington, D.C. 1976.
2. D. L. Brink, J. A. Charley, G. W. Faltico, and J. F. Thomas, "The Pyrolysis-Gasification-Combustion Process, Energy Considerations and Overall Processing," Thermal Uses and Properties of Carbohydrates and Lignins, F. Shafizadeh, K. V. Sarkanen, and D. A. Tillman, eds. Academic Press, 1976.

3. D. L. Brink and J. F. Thomas, "Pyrolysis-Gasification-Combustion I. A Recovery System for Pulping Liquor," in Alkaline Pulping Conference Preprints, September 15-18, 1974, Seattle, TAPPI, Atlanta, Ga., 1974. D. L. Brink and J. F. Thomas, "Pyrolysis-Gasification-Combustion: Design of an Experimental Unit Used for Study of Heat and Chemical Recoveries from Pulping Liquors," TAPPI 58(4), April, 1975.
4. R. T. Williams and D. L. Brink, "Pyrolysis-Combustion," Comprehensive Studies of Solid Wastes Management, Final Report, J. M. McFarland, D. L. Brink, C. R. Glassey, S. A. Klein, P. H. McGauhey, and C. G. Golueke, eds. SERL Report 72-3, University of California, Berkeley, May 1972.
5. D. L. Brink, J. F. Thomas, and K. H. Jones, "Malodorous Products from the Combustion of Kraft Black Liquor III. A Rationale for Controlling Odors." TAPPI 53(5), May, 1970.
6. D. L. Brink, "Pyrolysis-Gasification-Combustion: A Process for Utilization of Plant Material," Applied Polymer Symposium, 28(3). T.E. Timell, ed. John Wiley and Sons, 1976.
7. D. L. Brink, G. W. Faltico, and J. F. Thomas, "The University of California Pyrolysis-Gasification-Combustion Process," Forum on Kraft Recovery Alternatives. The Institute of Paper Chemistry, Appleton, Wis., 1976.
8. D. L. Brink, G. W. Faltico, and J. F. Thomas, "Pyrolysis-Gasification-Combustion. Feasibility in Pulping Recovery Systems - The First Stage - Second Stage Reactor as a Production Unit," Alkaline Pulping Conference, Sept. 13-15, Dallas, Tex. TAPPI, 1976.
9. D. L. Brink, S. Y. Lin, and J. F. Thomas, "Pyrolysis-Gasification-Combustion. Feasibility in Pulping Recovery Systems - The First Stage Reactor." Alkaline Pulping Conference, October 27-29, Williamsburg, Va. TAPPI, 1976.

10. J. K. Salisbury, Kent's Mechanical Engineers' Handbook, 12th Edition, Power Volume. J. Wiley and Sons, Dec. 1954.

11. R. H. Perry and C. H. Chilton, eds., Chemical Engineers' Handbook, 5th Edition. McGraw Hill Book Co., 1973.

WOOD OIL FROM PYROLYSIS OF PINE BARK-SAWDUST MIXTURE*

J. A. Knight, D. R. Hurst, and L. W. Elston

Engineering Experiment Station
Georgia Institute of Technology
Atlanta, Georgia

I. INTRODUCTION

Pyrolysis has received considerable attention in the last several years as a method for converting waste materials--agricultural, silvicultural, and municipal refuse--into useful products, particularly fuels; and at the same time, serving as a satisfactory disposal method. Pyrolysis is now receiving consideration as a process for the conversion of biomass from biomass plantations into useful products, particularly fuels [1]. Pyrolysis of lignocellulosic materials produces a char, condensible organic substance, water, and noncondensible gases. The char and condensible organic liquid can be utilized as clean burning fuels, and the noncondensed gas is a low Btu fuel that must be used on site.

* Supported in part by E.P.A. Grant No. R804 416 010.

Workers at the Engineering Experiment Station (EES) have found that pyrolysis is readily adaptable for the conversion of a variety of cellulosic and lignocellulosic materials into useful fuels and other products. During the past eight years, a steady flow, low temperature pyrolysis system has been developed which involves processing of the feed material in a continuous operation in a porous, vertical bed [2]. The Tech-Air Corporation, Atlanta, licensee for the process, has successfully operated a 50 dry ton/day field demonstration pyrolysis facility at a lumber yard at Cordele, Georgia. The feed material utilized at this facility is a mixture of pine bark and sawdust.

The physical and chemical characteristics obtained from the pyrolysis of various types of waste materials are needed for the proper evaluation and utilization of these oils. These data are also useful in establishing the pyrolysis operating conditions for production of the most suitable oils. There are many available testing and characterization procedures that have been developed for petroleum, vegetable, and other oils. Some of these, or modifications, will prove useful for oils obtained by pyrolysis of lignocellulosic wastes. The data presented below were obtained on wood oils produced in the Tech-Air facility.

II. WOOD OIL SAMPLES

Approximately 110 gallons of oil for this study were collected on July 15, 1976, at the facility while it was operating in a steady state mode. Approximately equal quantities of oil were obtained from the air cooled condenser and the draft fan, which is located between the condenser and the after-burner for the noncondensed gases. Samples of both oils were stored at ambient temperature and 0°C for future characterization and use. A representative sample of the feed material, a mixture of pine bark-sawdust, had the chararacteristics and properties as listed in Table I, and the results of the characterization and analyses of representative samples of both oils are given in Table II.

Certain properties were redetermined on samples of the oil, which had been stored at 0°C and ambient temperatures for approximately eight months. These results, along with the initial values, are given in Table III.

One significant use of these oils is as a fuel. Therefore, it is of interest to compare some typical properties of the condenser and draft fan wood oils with #2 and #6 fuel oils. These values are given in Table IV. Because of the greater densities of the wood oils, the heating values of the wood oils when compared on a volume basis are larger percentages of the heating values of the fuels oils than when compared on a weight basis. The very low sulfur content of the wood oils is a significant property for the utilization of these materials as fuels. Proper blending of wood oils with high sulfur fuel oils can serve as a means of reducing the overall sulfur content of the combined fuel, and consequently, sulfur emissions.

TABLE I. Properties of Pine Bark-Sawdust Feed Material

Property	Result	Method
Pine bark / Pine sawdust	70% / 30%	Microseparation by visual means
Bulk density	21.3 kg/cu m 13.3 lbs/cu ft	
Moisture	10.3%	ASTM D-1762-64
Ash (weight %)	1.3%	ASTM D-1762-64
Acid insoluble ash (weight %)	<0.1%	
Heating value (dry basis)	5061 cal/g 9109 Btu/lb	ASTM D-240-74

TABLE II. Properties of Wood Oils From Tech-Air 50 Dry Ton/Day Facility

Property	Condenser Oil	Draft Fan Oil	Method
Density	1.1415 g/ml 9.525 lbs/gal	1.1075 g/ml 9.242 lbs/gal	-
Water content (weight %)	14.0%	10.4%	ASTM D 95-70
Heating value (wet basis)	5,056 cal/g 9,100 Btu/lb	5,883 cal/g 10,590 Btu/lb	ASTM D 240-64
pH	2.9	3.3	5% oil dispersed in water
Acid number	75 mg KOH/g	31 mg KOH/g	ASTM D-664-58
Flash point	111°C 233°F	121°C 240°F	ASTM D-93-73
Filterable solids (weight %)	0.3%	0.4%	Acetone insoluble
Copper strip corrosion	1	1	Classification ASTM D-130-7
Sulfur (weight %)	<0.01%	<0.01%	ASTM D-129-64
Pour point	26.7°C 80°F	26.7°C 80°F	ASTM D-97-66
Ash (weight %)	0.08%	0.03%	-
Distillation			ASTM D-86 Group 3
First drop	98°C	101°C	
10% point	103°C	105°C	
48% endpoint	NA	265°C	
53% endpoint	282°C	NA	
Solubility (weight %)			
Acetone	99.6%	99.6%	
Methylene chloride	93.5%	97.8%	
Toluene	Slightly	Slightly	
Hexane	Slightly	Slightly	
Elemental analysis (weight %)			
Carbon	51.2	65.6	
Hydrogen	7.6	7.8	
Nitrogen	0.8	0.9	

TABLE III. Variation of Oil Properties Over Eight Months Period

Property	Condenser oil		
	Initial value	Stored eight months	
		0°C	Ambient temperature
Water content (weight %)	14.0%	20.5%	24.1%
Heating value (wet basis)	5,056 cal/g 9,100 Btu/lb	5,444 cal/g 9,800 Btu/lb	5,106 cal/g 9,190 Btu/lb
Acid number	75 mg KOH/g	87 mg KOH/g	89 mg KOH/g
Viscosity[a]	275 cP	350 cP	175 cP
pH	2.6	3.4	2.9

Property	Draft fan oil		
	Initial Value	Stored eight months	
		0°C	Ambient temperature
Water content (weight %)	10.4%	15.5%	12.7%
Heating value (wet basis)	5,883 cal/g 10,590 Btu/lb	5,922 cal/g 10,660 Btu/lb	5,939 cal/g 10,690 Btu/lb
Acid number	31 mg KOH/g	71 mg KOH/g	60 mg KOH/g
Viscosity[a]	233 cP	79 cP	475 cP
pH	3.3	3.1	3.0

a. Determined with Brookfield Viscosimeter, Model LV with Thermosel system at 25°C at 60r/min.

TABLE IV. Typical Properties of Wood Oils and Fuel Oils

	Wood oils[a]		Fuel oils[b]	
Property	Condenser	Draft fan	#2	#6
Water content, %	14	10.4	Trace	2
Btu/lb	9,100	10,590	19,630	18,590
Btu/gal	86,700	97,850	139,400	148,900
Density, g/ml	1.142	1.108	0.851	0.960
Lb/gal	9.53	9.25	7.10	8.01
Pour point	80°F	80°F	0°max	65-85°F
Flash point	233°F	240°F	100°F min	150°F
Viscosity, cP[c]	225	233	20	2262
Elemental analysis				
Carbon %	51.2	65.6	86.1	87.0
Hydrogen %	7.6	7.8	13.2	11.7
Nitrogen	0.8	0.9	-	-
Sulfur %	<0.01	<0.01	0.6-0.8	0.9-2.3

a. Values obtained on oils with moisture content as reported.

b. Values for fuel oils are considered typical. Sulfur will vary depending origin of oil. Ref.: North American Combustion Handbook, 1st ed., North American Mfg. Co., Cleveland, Ohio, 1952.

c. Determined with Brookfield Viscosimeter, Model LV with Thermosel system at 25°C at 60r/min.

III. VISCOSITY

The viscosity of liquids and its change with temperature is a significant property, particularly with liquids that will be handled by pumping. The viscosity values for the wood oils in this study were determined with a Brookfield viscosimeter, Model LV, with Thermosel system. The viscosity versus temperature was determined for both the condenser and draft fan oils initially and on samples which had been stored at 0°C and ambient temperature for approximately eight months. These viscosity curves are given in Figs. 1 and 2. The viscosity versus temperatures curves of samples of both oils which had been vacuum stripped for removal of water, as described in Section IV E, are given in Figs. 3 and 4. In order to determine the effect of prolonged heat upon the viscosity of condenser oil, samples of sealed oil were heated at 110°C for different time periods, and the viscosity was then determined for each sample. These data are presented in Fig. 5. For comparison, the viscosities of the condenser oil and #2 and #6 fuel oils are presented in Fig. 6.

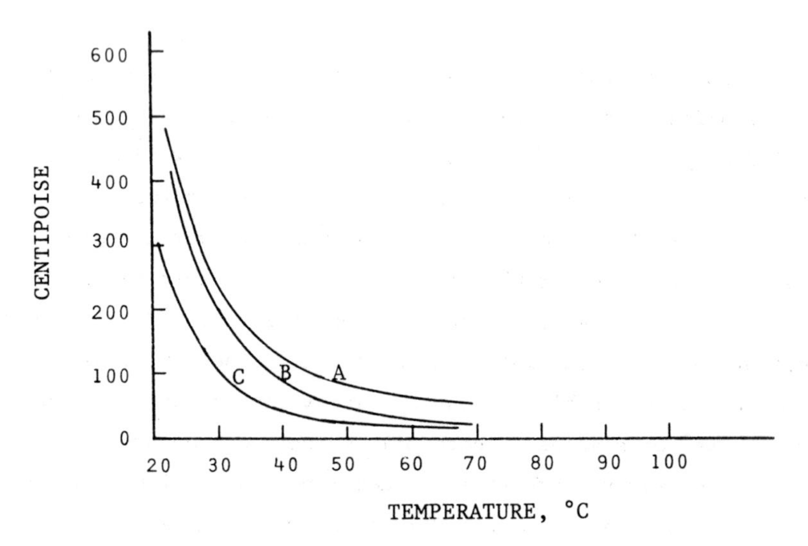

FIGURE 1. Viscosity of condenser oil: (A) Initial viscosity curve; (B) sample stored at 0°C for eight months; and (C) sample stored at ambient temperature for eight months.

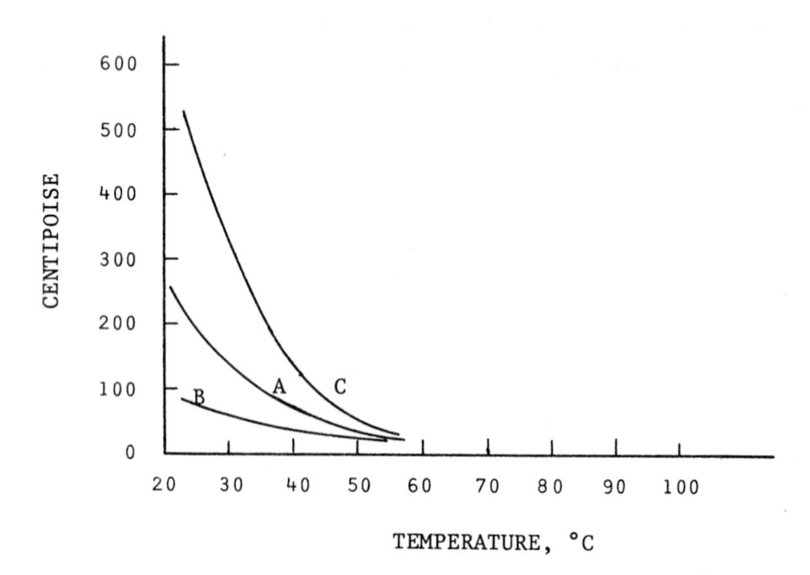

FIGURE 2. Viscosity of draft fan oil: (A) Initial viscosity curve; (B) sample stored at 0°C for eight months; and (C) sample stored at ambient temperature for eight months.

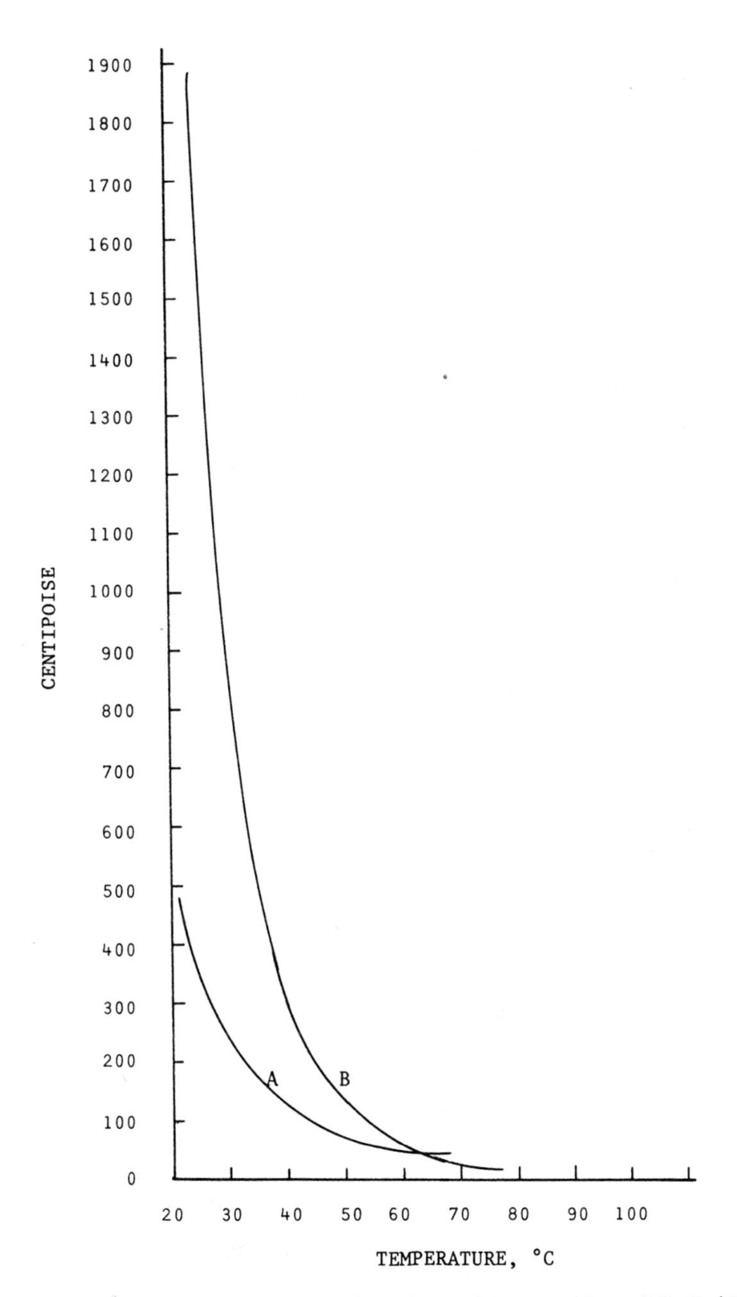

FIGURE 3. Vacuum stripped condenser oil: (A) Initial viscosity curve, and (B) vacuum stripped viscosity curve.

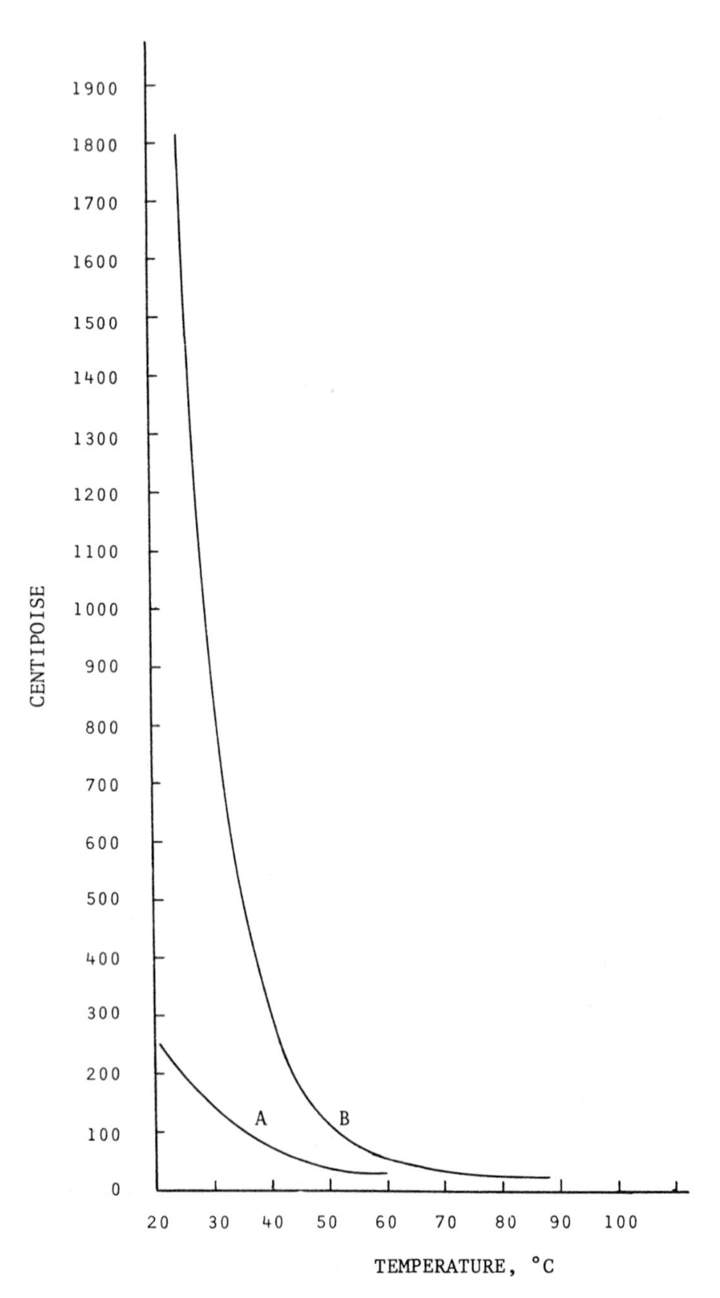

FIGURE 4. Vacuum stripped draft fan oil: (A) Initial viscosity curve, and (B) vacuum stripped viscosity curve.

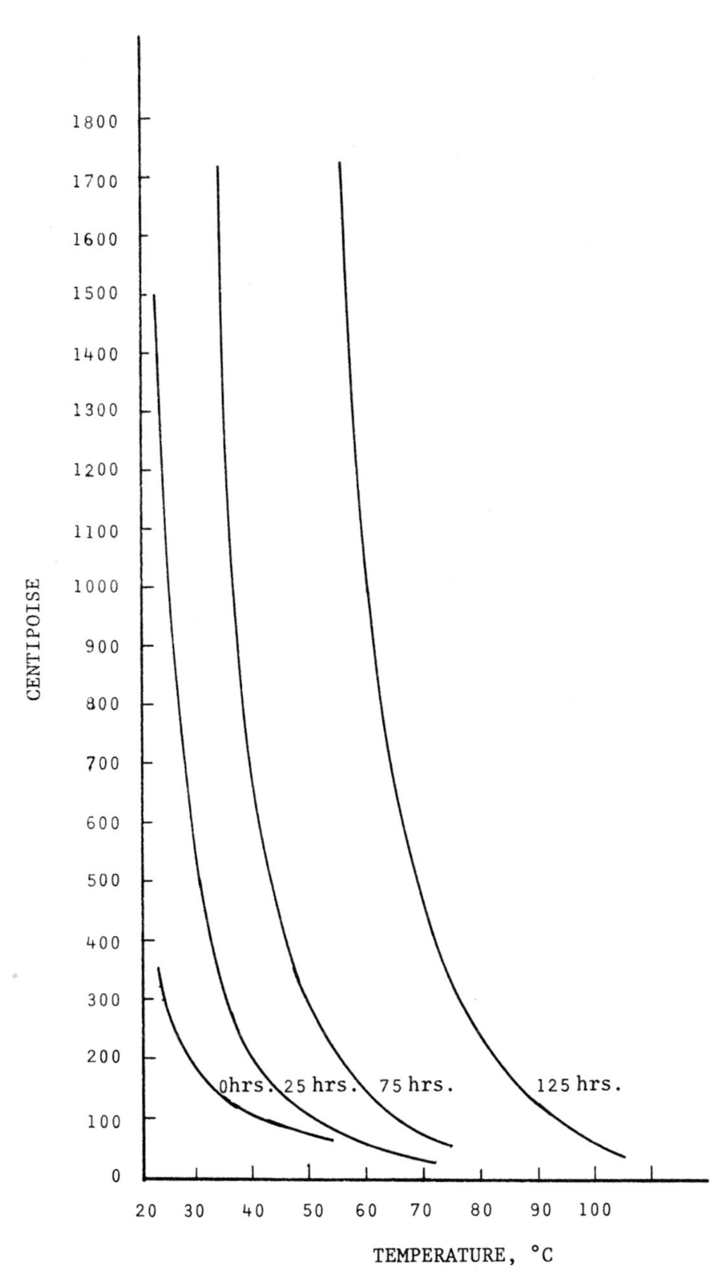

FIGURE 5. Effect of heating condenser oil at 110°C for different time periods on viscosity.

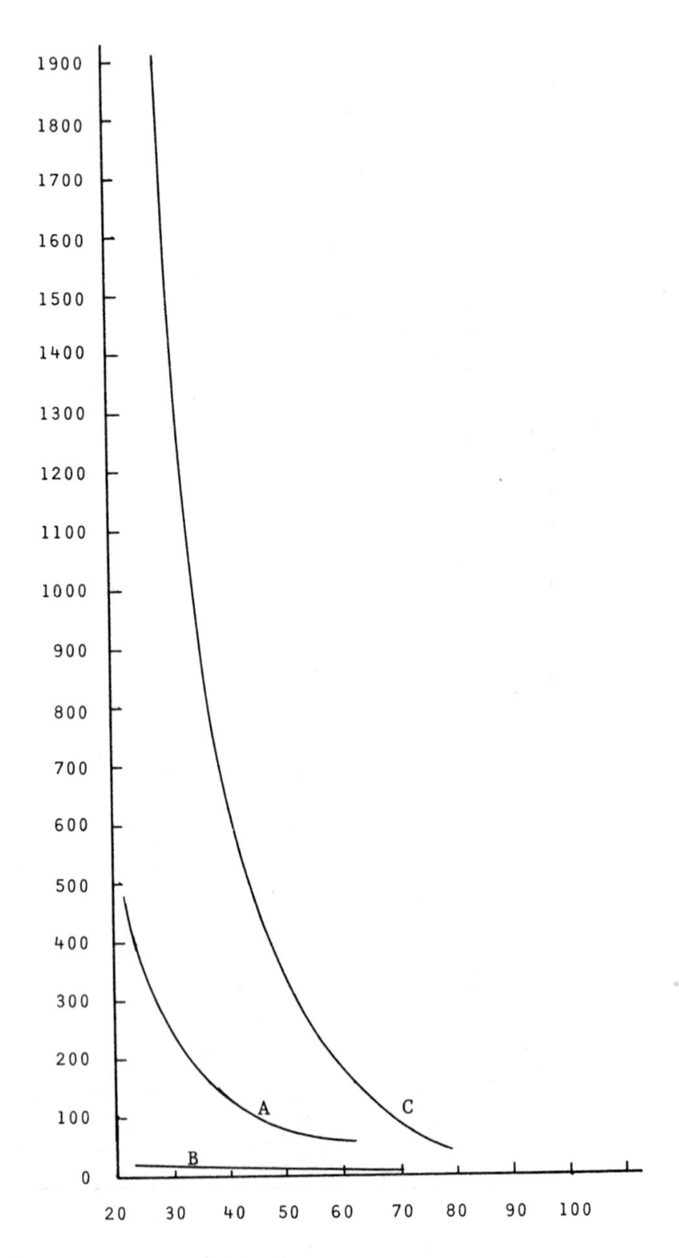

FIGURE 6. Viscosity curves for condenser oil (initial) and #2 and #6 fuel oils: (A) Condenser oil; (B) #2 fuel oil; and (C) #6 fuel oil.

IV. DISTILLATION

Distillation offers a possible method for processing wood oil to produce more desirable and useful products. Therefore, samples of the condenser wood oil were distilled under a variety of conditions.

A. Distillation at Atmospheric Pressure

A sample of condenser oil was distilled at atmospheric pressure, and three distillate phases were obtained--an aqueous phase, a heavy oil phase and a light oil phase. Most of the aqueous phase was obtained at 95°-125°C. Between 125°-190°C, the distillate was mostly heavy oil with some water and light oil. Between 190°-230°C, the distillate was mainly light oil. After approximately 68% of the sample had been collected as distillates, the head temperature began to drop, and the distillation was discontinued. The distillation data are presented in Table V.

TABLE V. Atmospheric Distillation Data

Fraction	*Head Temperature, °C*	*Yield, %*
Aqueous phase[a]	*95°-125°*	*20.7*
Heavy oil	*105°-190°*	*28.4*
Light oil	*190°-230°*	*19.1*
Residue[b]		*25.0*
Total		*93.2*

a. Contained 6.2% organic material as determined by gas chromatography.

b. The residue solidified on cooling to room temperature.

The analytical data on the oil phases and the residue are presented in Table VI.

TABLE VI. Analytical Data--Atmospheric Distillation Fractions

Property	Light oil	Heavy oil	Residue
Density, g/ml	0.9988	1.0423	-
pH	3.5	2.9	-
Acid No., mg KOH/g	52	17.8	-
Elemental analysis			
Carbon %	67.4	74.0	60.1
Hydrogen %	7.5	8.2	2.9
Nitrogen %	0.1	0.7	0.8

B. Vacuum Distillation

A sample of condenser oil was distilled in a simple vacuum distillation apparatus at 0.2-0.4 mm Hg pressure with dry ice and liquid nitrogen traps for recovery of volatile material. The aqueous phase containing light organics was obtained before the head temperature reached 43°C. At 190°C head temperature, the distillation was discontinued as the material in the flask was becoming more viscous and appeared to be approaching a decomposition stage. The vacuum distillation data are presented in Table VII.

The oil fractions 1-4 were combined to obtain a sample of oil for fractionation under vacuum with a spinning band column. Analytical data on the combined oil fractions is given in Table VIII.

TABLE VII. Vacuum Distillation Data

Fraction	Head temperature, °C	Yield, %
Aqueous and light organics phase[a]	Ambient to 43°	23.1
1	43°-125°	17.6
2	125°-140°	2.1
3	140°-180°	18.5
4	180°-190°	2.6
Residue	-	32.4
Total		96.3

a. Contained 46% organic material as determined by G.C. gas chromatography

TABLE VIII. Analytical Data-Vacuum Distillation

Property	Combined fractions 1-4
Density, g/ml	1.091
pH	2.8
Acid No., mg KOH/g	59.6
Heating value, cal/g, Btu/lb	10,964
Elemental analysis	
Carbon, %	62.7
Hydrogen, %	7.7
Nitrogen, %	0.3

C. Vacuum Spinning Band Distillation

A sample of the combined oil fractions from the vacuum distillation (B above) was distilled in a Nester-Faust spinning band distillation column at 0.2-0.4 mm pressure with liquid nitrogen traps for recovery of volatile material. The distillation data are given in Table IX.

TABLE IX. Spinning Band Distillation Data

Fraction	*Head temperature, °C*	*Yield, %*
1	35-48	8.0
2	48-58	14.6
3	58-60	8.5
4	60-68	2.1
5	68-74	6.9
6	74-82	6.0
7	82-88	4.6
8	88-98	2.8
9	98-105	5.5
Flask residue	-	30.0
Trap material	-	7.8
Total		96.8

At a head temperature of 105°C, the distillation had slowed down considerably. The material in the flask was becoming more viscous. The distillation was discontinued. Upon cooling to ambient temperature, the residue in the flask was a very viscous, black tarry material. Additional data on some of the fractions are presented in Section V.

D. Steam Distillation

A sample of condenser oil, 1238 g (14.0% water), was steam distilled. An oily layer, 8.8 g, along with an aqueous layer of 385 g was collected using steam only at a temperature up to 105°C. At this point heat was applied to the flask in addition to steam injection. The distillation was continued up to 190°C. The data on the distillates and flask residue are given in Table X.

TABLE X. Steam Distillation Data

	Temperature		
Material	*Up to 105°C*	*105°-150°C*	*150°-190°C*
Aqueous phase	*385 g*	*420 g*	*279 g*
Water	*354 g*	*359 g*	*228 g*
Organic	*31 g*	*61 g*	*51 g*
Organic phase	*8.8 g*	*33.8 g*	*183 g*
Residue, organic	-	-	*468 g*

Of the original charge, 18.2% distilled as a separate organic phase, and 11.6% organic material was detected by gas chromatography in the aqueous layers. A recovery of only 78.6% was obtained, which indicates that perhaps more organic material was present in the aqueous distillates than was detected.

E. Vacuum Stripping of Water from Wood Oil

The condenser and draft fan wood oils obtained from the Tech-Air facility contained water which does not separate on standing. Samples of the oils without the water were needed for characterization. Vacuum stripping was found to be the most effective means of removing this water. Two experiments for removal of the water were carried out at 0.3 mm of mercury. Dry ice-acetone and liquid nitrogen traps were used to recover any water and volatile organics. In the first experiment, the flask was heated to 55°C to insure that the water would be removed. Based on the amount of volatile organics recovered, a second experiment was carried out in which the flask temperature was maintained at 25°C. The data are presented in Table XI.

TABLE XI. Vacuum Stripping of Water from Condenser Wood Oil

	Weight	
Phase	*Experiment 1 Heated to 55°C*	*Experiment 2 25°C*
Aqueous phase		
Water	*12.1*	*12.9*
Organic	*4.4*	*0.0*
Organic phase	*0.35*	*1.3*
"Dried" oil	*82.4*	*84.3*
Total	*99.25*	*98.5*

V. LIQUID CHROMATOGRAPHY

Liquid chromatography (L.C.) was selected as one analytical technique which could be used for characterization of wood oils and fractions obtained from them. L.C. appeared to be particularly suitable for these oils since L.C. is carried out at ambient temperature, is capable of high resolution of complex mixtures, and detection of components is nondestructive. The main initial objective of utilizing L.C. in our work with wood oils obtained by pyrolysis is to use a method for "fingerprinting" the raw oils and fractions obtained from them for comparison and correlation. From preliminary experimental work, the conditions for obtaining survey L.C. chromatograms are as follows: Partisil ODS 5μ column; water-acetonitrile solvent system with 10 to 100% acetonitrile solvent gradient; and the uv detector set at 254 nm.

Representative chromatograms of raw condenser and draft fan oils are given in Fig. 7 and 8, respectively; of the distillate from simple vacuum distillation of condenser oil, Fig. 9; of fractions 1, 5, and 9 from the vacuum distillation of the oil distillate from the simple distillation of condenser oil, Figs. 10, 11, and 12, respectively; of vacuum stripped condenser oil, Fig. 13; and of organic and aqueous phases from steam distillation of condenser oil, Figs. 14 and 15.

VI. RESULTS AND DISCUSSION

The wood oils for this study were obtained from the demonstration facility of the Tech-Air Corporation. The feed material was a mixture of pine bark-sawdust in a ratio of approximately 70:30. The pyrolytic converter is a vertical shaft type reactor in which the feed material is fed in at the top giving a vertical, porous bed of the feed material. Char is removed at the bottom and the gaseous phase passes up through the porous bed of feed material into the off-gas system. The air cooled condenser is operated to condense an organic phase which contains 15% or less water. A second organic phase is obtained from the draft fan located between the condenser and the after-burner for the non-condensed gases.

The condenser and draft fan wood oils obtained for this study were free flowing liquids and had a dark brownish color and a burnt odor. The water in the oil is well dispersed or emulsified and does not separate into an aqueous phase on standing. The densities of the oils are greater than water. The oils are acidic and exhibit some corrosive properties. The heating

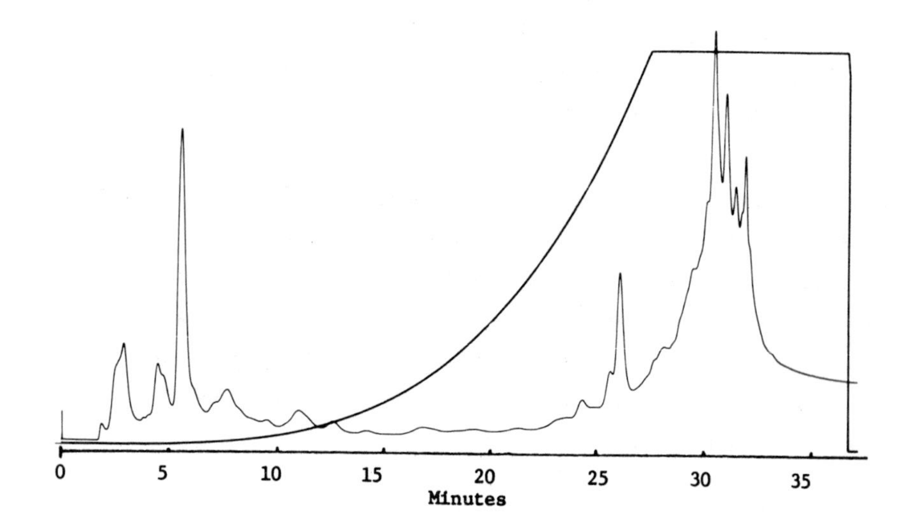

FIGURE 7. Survey liquid chromatogram of raw condenser oil.

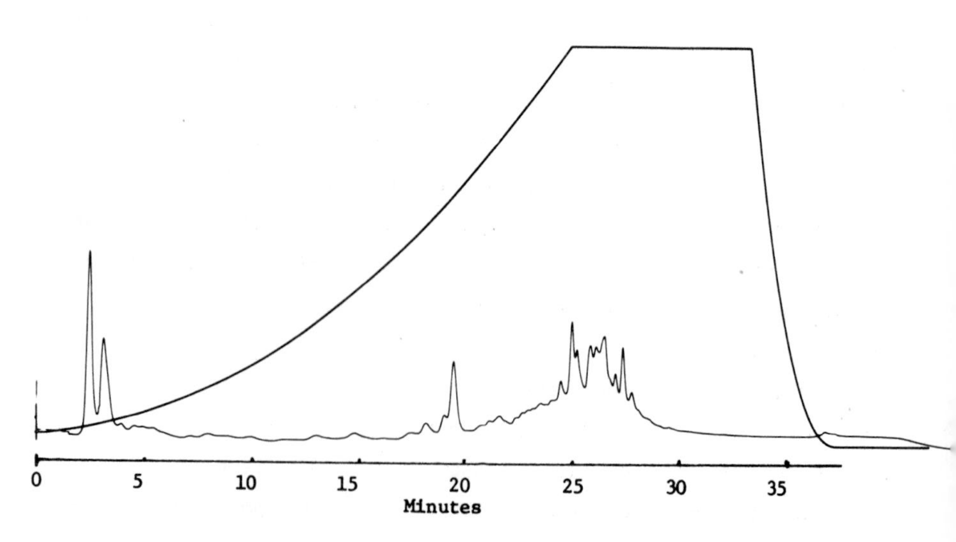

FIGURE 8. Survey liquid chromatogram of draft fan oil.

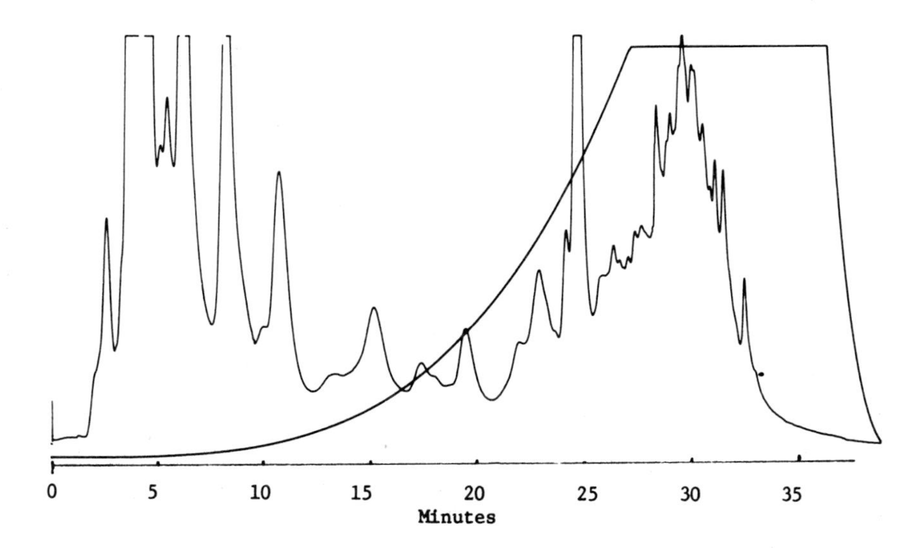

FIGURE 9. Survey liquid chromatogram of combined fractions from vacuum distillation.

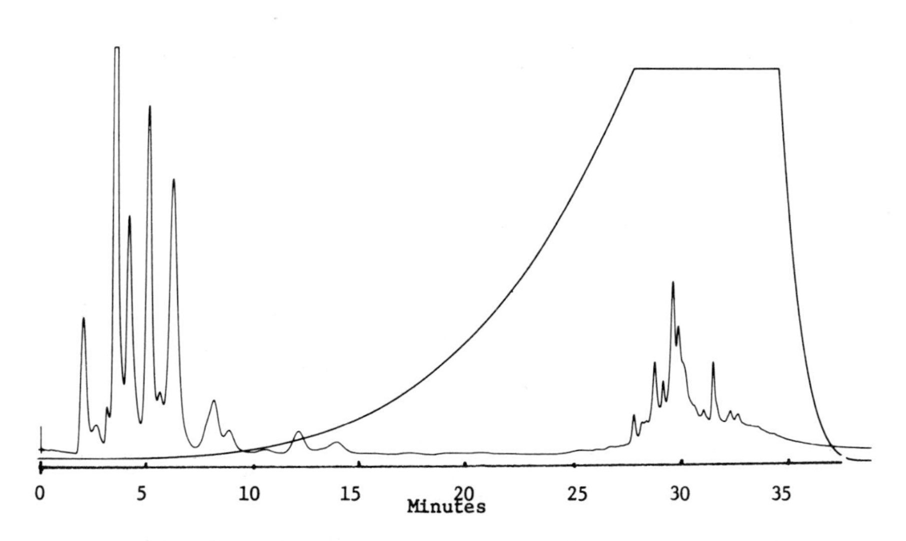

FIGURE 10. Survey liquid chromatogram of spinning band 1 fraction.

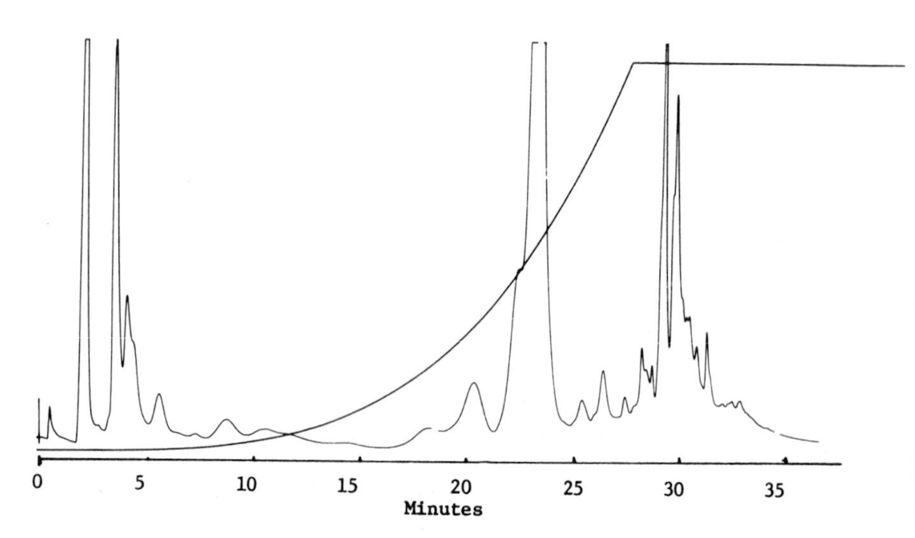

FIGURE 11. Survey liquid chromatogram of spinning band 5 fraction.

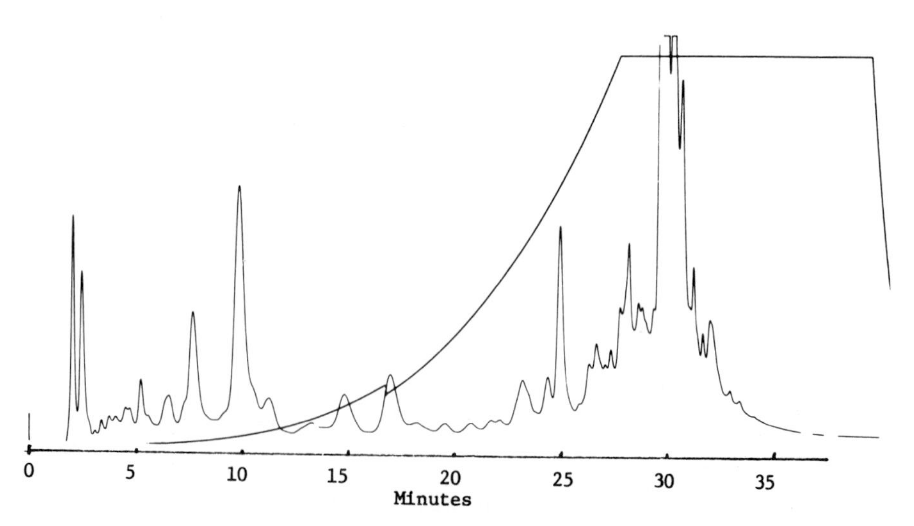

FIGURE 12. Survey liquid chromatogram of spinning band 9 fraction.

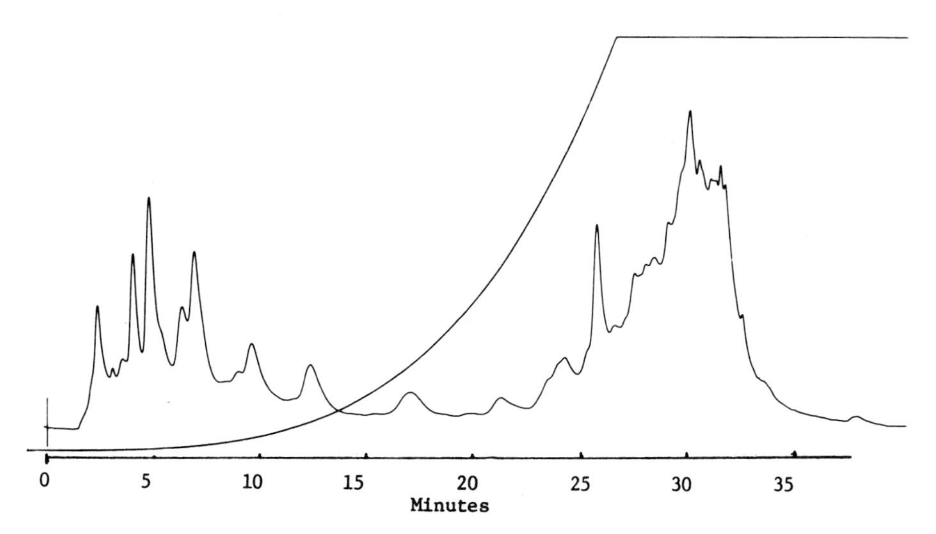

FIGURE 13. Survey liquid chromatogram of condenser oil vacuum stripped without heat.

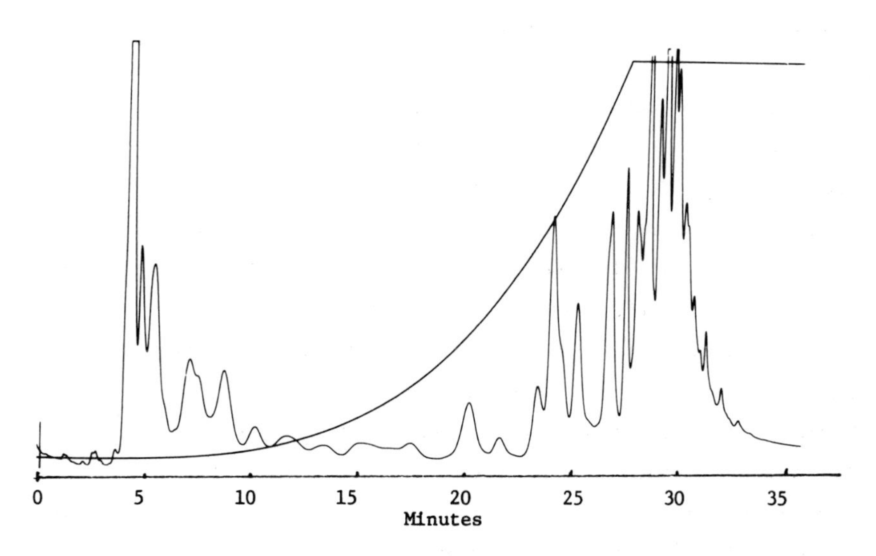

FIGURE 14. Survey liquid chromatogram of 100°-105°C organic layer from steam distillation.

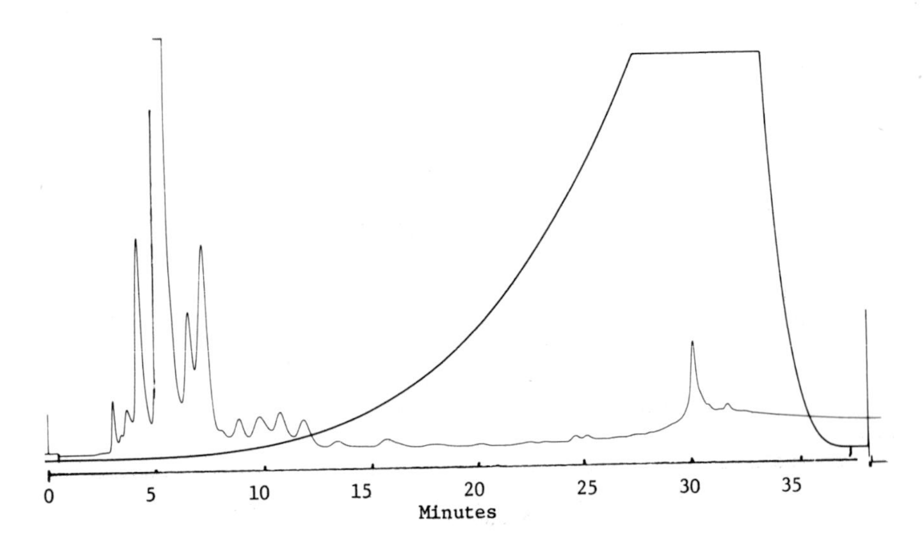

FIGURE 15. Survey liquid chromatogram of 100°-105°C aqueous phase from steam distillation.

values of the oils are less than those of fuel oils, since the wood oils contain a fair percentage of oxygen. The properties of samples of the condenser and draft fan oils which were stored for eight months showed some changes (Table III). However, these changes do not appear to be highly significant in the utilization of the oils as fuels.

The wood oils have greater densities than the #2 and #6 fuel oils. For this reason the heating values of the wood oils are a greater percentage of the heating values of the fuel oils when compared on a volume basis than when compared on a weight basis. The very low sulfur content of these wood oils and similar oils obtained from different lignocellulosic materials is very significant. One utilization approach is blending or co-firing of these oils with sulfur containing fuels to reducing sulfur emissions at a given plant to acceptable levels.

The viscosity of oils is an important property in the handling of these materials. The viscosities of a number of wood oil samples and #2 and #6 fuel oils were determined and are given in Figs. 1 through 6. The viscosities of the wood oils changed some on storage at 0°C and ambient temperature over an eight month period. These changes should not have any great effect on the handling characteristics of the oils. The vacuum stripped oils had much higher viscosities at the lower temperatures, but approached the viscosities of the fresh oils at the higher temperatures. These results are essentially what would be expected since the vacuum stripping removes the water in the oils. Sealed samples of the condenser oil were heated at 110°C for different time periods, and the viscosity curves are given in Fig. 5. The curves show that the viscosity of the heated samples increases markedly with time. The curve for the sample heated 75 hours is very similar to the viscosity curve for #6 fuel oil. These data show that prolonged heating of oils obtained by pyrolysis must be considered in the materials handling of the oils.

A number of distillation experiments were carried out with the condenser wood oil. The data from these initial experiments indicate that distillation is a potential processing method for obtaining fractions which could have greater utility and value than as a fuel. Additional study is needed on the chemical characteristics of the distillates, however, to find specific uses for them. The wood oils are heat sensitive, and in both atmospheric and vacuum distillation at heat temperatures of 190°-200°C, the material in the flask becomes more viscous and decomposes if heating is continued. The distillate oil from vacuum distillation of condenser oil exhibits the same behavior as a sample fractionated in a spinning band column at 0.2-0.4 mm pressure.

Liquid chromatography is an excellent analytical tool for characterization of oils produced by pyrolysis of lignocellulosic materials. Chromatograms on the condenser and draft fan oils and different distillate oils obtained by different distillation techniques show that the oils contain a large number of components. An examination of Figs. 10, 11, and 12 for three different fractions of oil obtained from the vacuum fractionation of condenser distillate oil shows that each of these fractions contains a large number of components.

VII. SUMMARY

The condenser and draft for oils produced by pyrolysis of a pine bark-sawdust mixture has been characterized and tested by a variety of methods. The oils, as produced, contain approximately 15% or less water, which is well dispersed. The heat values of the oils vary from about 60 to 70% of the heating values of fuel oils on a volume basis. The oils contain essentially no sulfur, and therefore do not create any sulfur emission problems when burned as a fuel. The tests for oil samples stored eight months at 0°C and ambient temperature indicate that no significant changes occurred which would affect the use of the oil as a fuel. The oils are acidic and exhibit some corrosive properties. The viscosity of the oils increased with prolonged heating at 110°C, and this factor would have to be considered in the handling of the oil. Also, the oils start to degrade and decompose when heated to approximately 200°C. The oils contain a fair percentage of oxygen mixtures with a large number of components as shown by liquid chromatographic analysis. It is reasonable to assume that many of the components are oxygen containing compounds.

REFERENCES

1. "Conference on Silviculture Plantations," The Mitre Corporation/Metrek Division and Georgia-Pacific Corporation, Reston, Va., February 15-16, 1977.

2. J. A. Knight, M. D. Bowen, and K. R. Purdy, "Pyrolysis--A Method for Conversion of Forestry Wastes to Useful Fuels," Presented at poster session, Conference-Energy and Wood Products Industry, Forest Products Research Society, Atlanta, Ga., November 15-17, 1976.

PROSPECTS FOR CO-GENERATION OF STEAM AND POWER IN THE FOREST PRODUCTS INDUSTRY

L. N. Johanson and K. V. Sarkanen

Department of Chemical Engineering
University of Washington
Seattle, Washington

I. INTRODUCTION

The oil embargo and subsequent tripling and quadrupling of energy costs has brought the term "co-generation" into familiar use in the United States. This means simply the combining of electric power generation with steam generation for other uses, at a single site. The term was used by President Carter in his 1977 address to the nation on U. S. Energy Policy, when it was cited as a prime possibility for early energy savings in U. S. industry. On other continents where high energy costs prevail, however, co-generation has been known and utilized for some time on a restricted basis. For example in the pulp and paper industry of Scandinavia, the technique has been in use for a number of years, under the terminology "By-product Power." In Finland, by-product power was utilized at approximately 50% of the possible level in 1960 and in 1974 was at about 80% of possible level. The corresponding 1974 levels in this industry in Sweden and the United States were approximately 50% and 25%, respectively [1].

In conventional central-station electric power generation in the United States, exhaust steam from the last turbine stage is condensed using the coolest water available from cooling towers or natural sources. This not only results in thermal pollution problems in many instances, but limits the conversion efficiency to 39% or 40% in a good fossil fuel fired plant today. When electric power can be generated with turbines expanding steam to the pressure required for process use, overall utilization of fuel can range to 88% or 89%, and incremental efficiency for power generation is about 85% [2].

Co-generation, or "by-product power," of course, is feasible only where the combined need exists for both steam and power. The pulp and paper industry is one of the most promising for such application, because of the widespread need for process steam at various pressures, and concomitant power requirements. Two specific examples are in the production of bleached-kraft pulp and of newsprint. In the former, heat requirements are 14 million Btu per ton of pulp, and power requirements 720 KW hr per ton. By means of co-generation, 1120 KW hr of electrical energy per ton of pulp can be produced, resulting in a salable surplus of 400 KW hr per ton. Newsprint requires less steam but more power, and the corresponding figures are heat requirements of 7 MM Btu per ton, power requirements of 360 KW hr per ton, and co-generation potential of 700 KW hr per ton, resulting in a potential net power requirement of 660 KW hr per ton of pulp [1].

The possibility of co-generation also is attractive in the pulp and paper industry in terms of the total U.S. potential and its contribution to overall energy conservation. The industry in 1972 produced 64 million tons of product, consumed 1.5×10^{15} Btu (1.5 Quads) of purchased fuels and 30 billion KW hr of purchased electric power from utilities. An additional

30 billion KW hr of electricity was generated internally [3]. If maximum use of the industry's co-generation potential could be realized, it could become a seller, rather than a buyer of electrical energy, with a potential energy saving of about 0.2×10^{15} Btu per year. This could be extended to a saving of about 0.75×10^{15} Btu per year with the incorporation of gas-turbine driven generators in combination with exhaust heat process steam boilers.

The pulp and paper industry, or more generally, the forest products industry, has an additional opportunity to help alleviate the energy crisis through the use of renewable fuels in place of fossil fuels. This is done already to a considerable extent through steam generation by burning of bark and wood wastes, and spent liquor evaporation and combustion in kraft pulping and so-called soluble base sulfite pulping systems, as has been pointed out in Mr. Tillman's paper. As fossil fuel costs increase, the incentive to displace such fuels in processes where they are used now increases, but this requires more complete utilization of wood wastes, particularly at the point of harvest, and at smaller, more isolated lumber and wood-use locations. The trends may be illustrated by reference to typical Southern Pine Stands [4]. In the early 1960s, waste constituted 35% of the stand, with lumber coming exclusively from 9-inch and larger logs, and pulp chips from logs 4-inch and larger. By 1974 waste had decreased to 27%, chiefly by increased utilization of 4- to 9-inch logs for both lumber and pulp chips. So-called "whole tree utilization" of the same stand will allow using all but 7% of the stand, largely through uses of small-log waste together with limbs, tops, and stems as fuel. Such improved utilization is expected to take place in spite of the shutting down of many small log operations, and the trend to greater centralization of the lumber and pulp industries. The greater hauling distances that result are more than compensated for by greater efficiency of wood use on the larger, integrated scale.

As energy supplies become increasingly scarce and costly, it may be advantageous for this industry to extend its fuel sources by looking outside the forest to other waste materials, particularly municipal and agricultural wastes. One such study [5] is for the utilization of municipal wastes of Muskegon, Michigan in a suspension-fired boiler of a nearby pulp mill, to supplement its usual fuel sources. Mr. Greco's paper references a similar utilization scheme in Northumberland, New Hampshire.

II. ADVANTAGES OF COMBINING CO-GENERATION WITH ENERGY FROM RENEWABLE RESOURCES IN THE FOREST PRODUCTS INDUSTRY

It should be apparent from the foregoing discussion that there is an opportunity in the forest products to reduce energy costs both by increased utilization of wood wastes and by co-generation of steam and electric power. If these actions are combined and are practiced on an extensive scale within the industry, they can make a significant contribution to national efforts to reduce dependence on fossil fuels. The most obvious combination of facilities is a kraft pulp mill and bleach plant with a power boiler fueled with wood wastes, with exhaust steam from power generating turbines utilized in the pulp mill and bleach plant. This is essentially state-of-the-art practice in some Scandinavian mills. Greater potential and flexibility will be realized however, if the concept is expanded to combine an integrated lumber-plywood mill, a pulp mill-bleach plant and paper mill, with a power generation facility. The latter should have the flexibility to utilize various solid wastes and fossil fuels (coal, lignite) and power cycles giving high conversion efficiencies; and, at the same time, meet stringent air and water pollution control requirements. If this can be done, several additional advantages ensue. The whole tree utilization concept with optimizing of wood-end-use distribution becomes possible. Not only steam but power can be distributed economically to the required use points with reduction or elimination of the need for purchased power. Water requirements for the overall processes can be minimized through extended recirculation patterns throughout the complex. As the complex approaches "total recycle" of water, that entering as moisture in the wood may supply most or all of the makeup needs. Chemicals for pulping which are now recycled through recovery systems in the kraft process can be supplemented by sodium and potassium carbonates, and possibly sulfur chemicals as well, from fossil fuel combustion, to further reduce chemical makeup requirements.

The combustion system for wood and solid wastes must meet the air pollution requirements for sulfur compounds and for particulate matter. The former is no problem if wood refuse is burned. It may be a problem if wood waste is supplemented with coal. The particulate requirements have caused difficulty with older "hog-fuel" boilers, but can be met with modern, well-designed power boiler systems. The paper by Mr. Voss provides a discussion of modern combustion systems.

An alternative to direct combustion systems, however, is to utilize a pyrolysis-gasification system, generating a low Btu or medium Btu gas which may subsequently be used for power generation. Char and oils may or may not be produced as by-products, depending on the process. If such a gas is produced, an additional advantage of the integrated mill system would be the possibility of providing gaseous fuel for operation of the kraft mill lime kiln, as auxiliary fuel for the recovery furnace, and for kiln drying of plywood and lumber, thus further minimizing oil or gas fuel needs.

Several possibilities exist for pyrolysis-gasification systems. Some of these are discussed in the papers by Mr. Voss and Dr. Brink. Three concepts will be explored here. These three more promising schemes are the Lurgi-type rotary grate packed bed, the Purox-type slag-ash packed bed, and the fluidized bed. A version of the Lurgi-type packed bed has been described by the Battelle Pacific Northwest Labs [6] for use with solid wastes. In this scheme shown in Fig. 1, solid waste moves downward counterflow to an air-stream supply entering at the bottom. The fuel is dehydrated and pyrolyzed as it descends, and the generated gas, cleaned somewhat by the cold entering fuel, exits at the top of the retort. Ash is removed through a rotary grate and airlock system at the bottom. The Purox™ retort is a development of the Union Carbide Company for gasifying municipal waste. It has no interior moving parts, but utilizes downflow solid fuel and upflow gas as does the Lurgi-type retort [7]. The gas in this instance is oxygen rather than air, supplemented by steam if desired. The oxygen develops a high combustion zone temperature, which melts metal, glass, and ash components of the fuel. This slag is then water-quenched upon exit from the bottom of the retort. The use of oxygen rather than air results in a higher Btu product gas, and indeed this may be viewed as a "synthesis gas" suitable for chemicals production (methanol, ammonia) as well as fuel. The third type of pyrolyzer is a fluidized bed of shredded or chipped solid fuel through which air or oxygen plus recycle gas passes upward. In the specific configuration shown in Fig. 2, to be further discussed later, it is a concept only, but the fluidized bed has been proposed in several versions for coal gasification, and it is in use at present for the combustion of magnesia-base sulfite pulping liquor, for combustion of secondary sewage sludge, and for power generation by combustion of solid wastes, among other applications.

A major advantage of gasifying solid waste fuels followed by combustion of the gas, is that the exit gas from the pyrolyzer

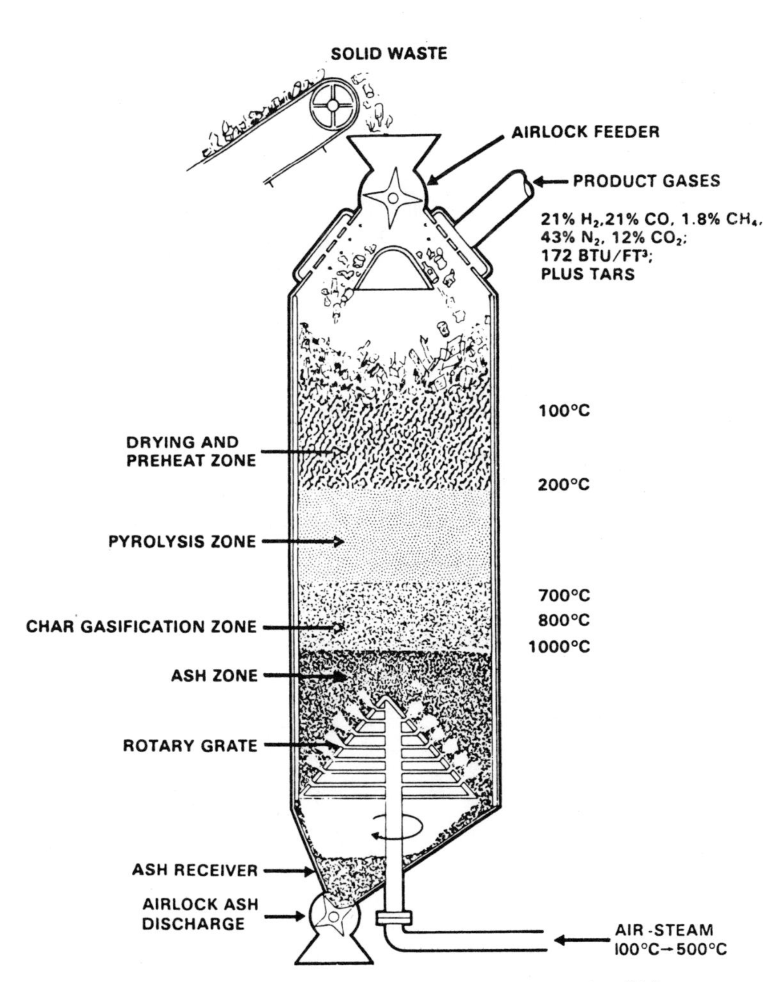

FIGURE 1. Moving bed waste pyrolysis unit [6].

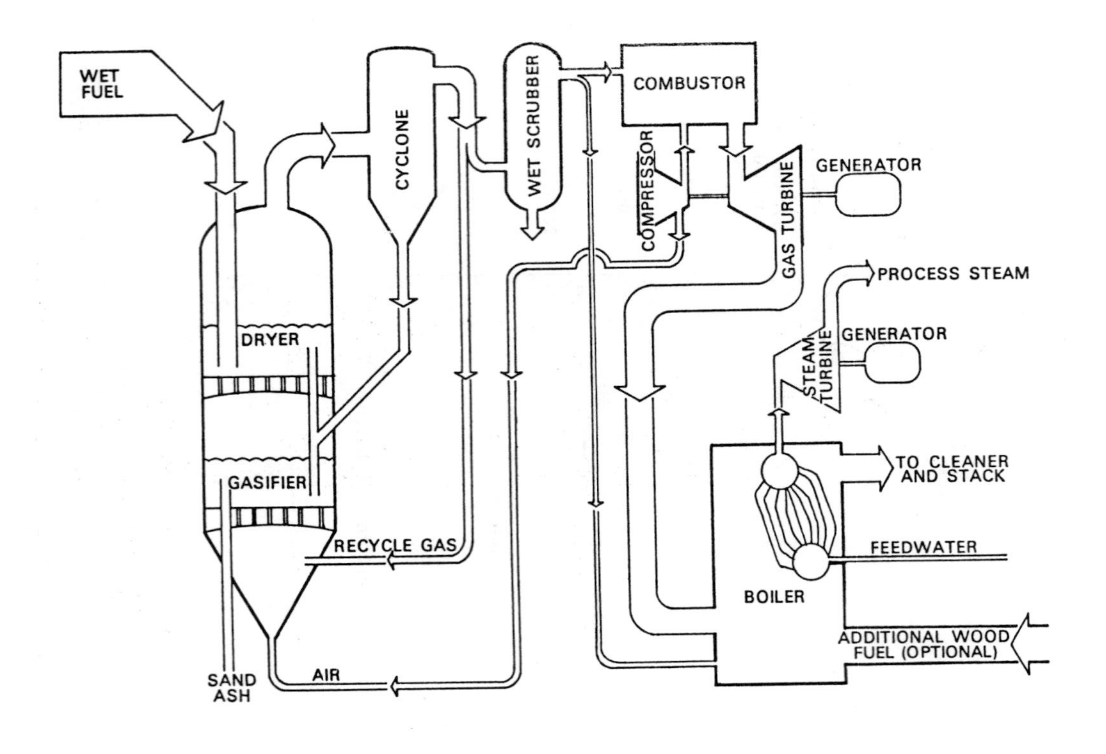

FIGURE 2. Fluidized bed pyrplysis-gasifier, incorporating a gas turbine power recovery cycle (reproduced from Fig. 14 of Ref. 9).

is smaller in quantity (about one fourth the volume on an equal-temperature basis) and is much cooler than the typical stack gas. Thus gas cleaning, particularly by scrubbing, is much less costly. Such a cleaning step can remove most of the water vapor, if desired, and there are then several options for the use of the resulting cleaned gas. In the present instance, such a gas-fuel supply fits in nicely with the concept of co-generation and forest industry integration as will be described more fully below.

III. PRESENT STATUS OF ENERGY USE IN THE FOREST PRODUCTS INDUSTRY

The lumber and plywood industry is in transition from the old methods of burning of some of the wood wastes without utilizing the energy ("teepee burners"), to complete utilization of all wood wastes in product end-uses or in steam or power generation. The latter situation is attained in only a few instances where scale of operation is sufficient and an adjacent or nearby pulp mill can utilize wood waste for pulp chips and steam production.

Recent increases in hydrocarbon fuel prices have prompted conversion of some fuel-using operations to wood-waste combustion with spectacular results. In a specific example in 1974-75 at Omak, Washington [8], replacement of propane gas fuel in two veneer dryers with double-vortex suspension type wood burners saved almost $500,000 per year in fuel costs assuming the wood to have no value. Alternatively, the wood value as fuel in terms of savings is approximately $50 per ton based upon propane prices of 30¢ per gallon. These prices are in the same range as present prices for quality pulp chips and fuel oil.

Such trends are encouraging and are taking place rapidly, but much remains to be done with respect to waste wood in the forest. Grantham [9] has estimated (1974) that of 69.2 million tons of annual timber harvest on the Pacific Coast, 20% remains behind as logging residue, 8.2% as sawmill residue, 11.4% is burned as fuel, 38.1% used as lumber and wood products, and 19.8% as pulp wood.

The pulp and paper industry annual fuel and power purchases for 1972 have been indicated earlier. The total fuel used is difficult to estimate since varying amounts of wood wastes, bark, etc., are utilized, in-house or purchased. Fuel derived from spent kraft liquors represent a tonnage about equal to the pulp produced, while some sulfite pulping processes utilize little or none of the spent liquors as fuel. With sufficient fuel and power cost incentives or other incentives to justify the capital expenditure, substantial overall energy savings could result from increased generation of power, especially by more complete use of bark and wood waste, which are now estimated to be only 50% utilized. Additional fuel savings could result from decreased levels of bleaching of pulp, from the use of waste paper as fuel rather than recycling by repulping (in effect trading new wood for fuel) and other substantial changes in pulping technology. Replacement of the chlorine-using stages of pulp bleaching with oxygen bleaching would reduce energy use through reduction in use of energy-intensive chlorine, and release as fuel the combustibles removed from the pulp in the bleaching stages.

Oxygen has been indicated as a possible replacement for air in solid waste gasifier systems, and also as a bleaching agent in place of chlorine. Also it has possible benefits as an oxidizing agent in the kraft black liquor recovery system as an odor-control measure, as a BOD reduction agent in place of aeration in biological waste treatment, and as a possible pulping agent for "oxygen pulping" processes now under development. Therefore, it may be of interest to consider the installation of a liquid-air oxygen plant in our integrated co-generation forest products industry complex.

IV. A CONCEPTUAL INTEGRATED WOOD PRODUCTS AND PULP AND PAPER COMPLEX WITH A GASIFICATION-POWER GENERATION SYSTEM

Combining those features so far discussed which lead to improved overall energy economy results in an integrated industrial complex as depicted schematically in Fig. 3a and b. A particular gasifier-power plant complex has been illustrated previously as Fig. 2. This combination was proposed by the present authors in a study of wood utilization for the U. S. Forest Service, and is reproduced here from Fig. 14 of reference [9] by John B. Grantham of the Forest Service. Alternate gasifiers, such as the Battelle solid-waste retort or the Union Carbide Purox™ gasifier also could be considered. In Fig. 2, if oxygen is available from a liquid air plant, or as "partially-spent" oxygen from oxygen bleaching, oxygen-pulping, or black liquor oxidation, it could replace, in whole or part, the air feed to the gasifier. This would reduce the size (or increase the capacity) of the gasifier cyclone and scrubbing systems, and result in a higher Btu gas. The latter would have increased usefulness as pipeline fuel to a lime kiln, veneer-drying kiln, etc., at some distance.

The larger part of the generated fuel is compressed for use in the gas-turbine combustor, and the remainder is sent to a conventional boiler for steam generation. Air would be compressed for use in the combustor whether or not oxygen is available, as the dilution effect of nitrogen is not a disadvantage in the gas turbine application. Note that pre-cleaning of the fuel gas allows this use of solid waste as an indirect fuel for gas turbine power generation. The gas turbine exhaust is still at a substantial temperature (800-1000°F) and is used, together with remaining fuel gas, in an auxiliary "waste heat" boiler to generate steam. This is used to generate additional power, with the exhaust from the last stage (and at intermediate pressures as well, if desired) utilized as process steam. The gasifier

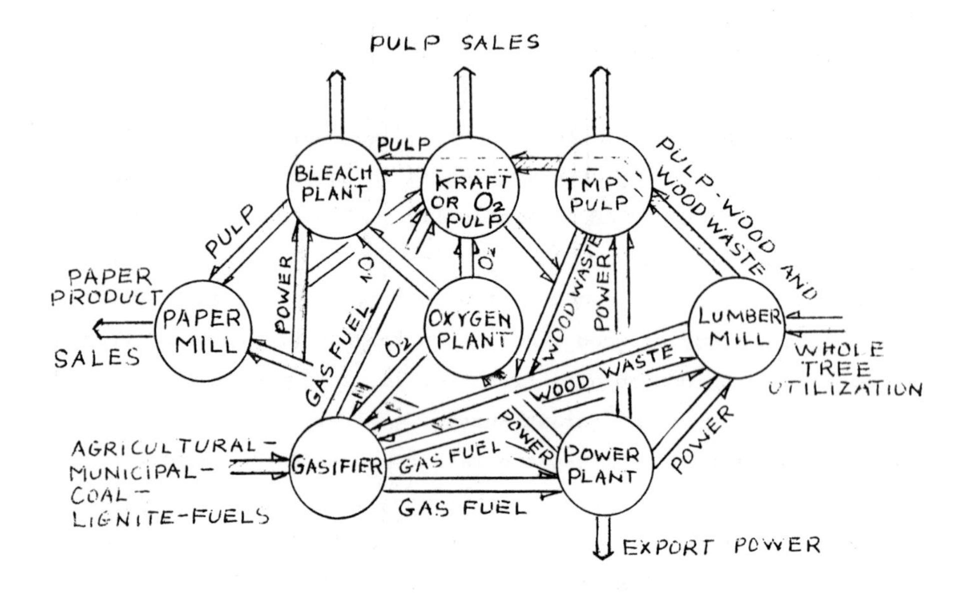

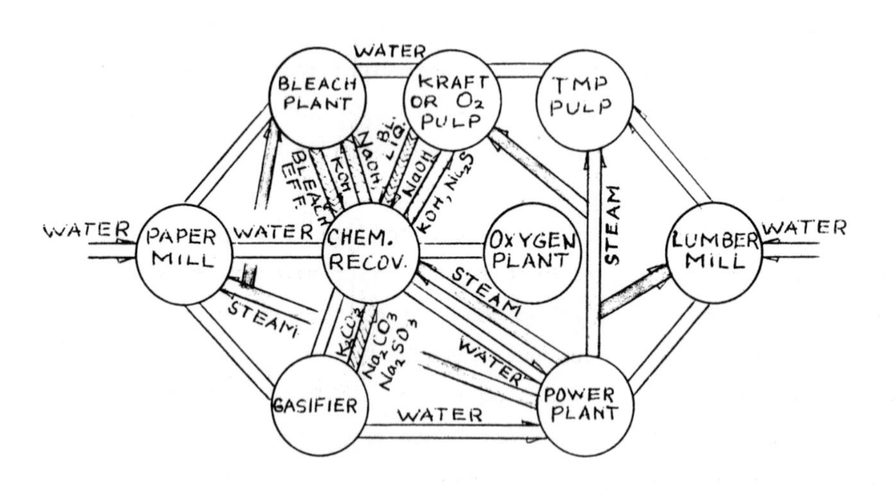

FIGURE 3. Co-generation combined with an integrated wood-processing complex, showing extensive material and energy relationships. (a) Raw materials, fuel and power input-outputs, (b) Water, steam and chemicals input-outputs.

shown is intended to operate at about 1600°F in the lower bed, and about 250 to 300°F in the upper bed. This provides for controlled predrying of the entering wood waste or solid waste. The uniform high temperature in the lower bed ensures complete conversion of organic compounds and carbon to largely hydrogen, carbon monoxide, water vapor, carbon dioxide and nitrogen. With the indicated temperatures, gasifying wood with air, the exit scrubbed gas is estimated to be 27% (volume) hydrogen, 31% (volume) carbon monoxide, and to have a higher heating value of about 190 Btu/scf.

Returning to Fig. 3, a raw materials, fuel, and power input-output diagram is shown for the overall complex in the top figure. A water, steam, and chemicals input-output diagram is in the lower figure. All but a few percent of the "whole tree" is brought to the lumber mill, from which remaining pulp wood and wood waste is sent to the gasifier for energy use, or as chips for kraft or thermomechanical pulping. Supplementary solid wastes also may be burned in the gasifier. If municipal wastes, these would probably be required on a year-round contract basis, with appropriate short-term storage. Pre-pelletizing and stabilizing of such wastes can be practiced to allow such short-term storage. If agricultural wastes are utilized, these would be supplied on a seasonal basis. In this case supplementary wood waste, coal, or lignite may augment the usual supply of wood waste from the lumber mill in periods of low supply of agricultural residues.

Gaseous fuel generated is used mainly in the power plant, but may be used to replace hydrocarbon fuels in the lime kiln, recovery boiler, and veneer or lumber drying kiln. Generated power is utilized in the lumber and pulp mills, bleach plant, oxygen plant and paper mill, but surplus may also be sold to a utility. Oxygen may be used for bleaching, black liquor oxidation, or oxygen pulping, but "waste" oxygen from the latter operations and additional generated oxygen may be used in the gasifier in place of air.

In the lower figure, steam from the power turbines (the co-generation feature) or steam boiler is utilized as motive power for pumps, etc., and as process steam throughout the complex. Water similarly is used throughout the complex as boiler feedwater, process and wash water and cooling water. In a modern, well-designed mill, much less net water is required than formerly, owing mainly to more stringent water pollution control requirements, and if utilization of condensate from evaporators and the gasifier scrubber is practiced, the water entering with the wet wood may be sufficient to operate the plant. However, supplementary water is arbitrarily shown entering the complex at the paper mill.

In the kraft pulping process, sodium hydroxide and sodium sulfide are used as pulping and bleaching chemicals. These are largely regenerated in the chemical recovery system. Smaller amounts of make-up sodium as sulfate or carbonate, and limestone and possibly sulfur are required, together with bleaching chemicals. The ash from wood waste combustion or gasification contains the mineral constituents of the tree plus bark, dirt, and salt contamination. It is well known since the days of primitive soap-making that potassium carbonate can be easily leached from such ash, and could serve readily as pulping and bleaching chemical, replacing makeup sodium salts. Alternatively, this ash can be returned to the forest as fertilizer. Sulfur also could be derived from sulfur-bearing fuels such as coal or municipal wastes, recovered as sodium bisulfite from the gasifier scrubber and used as makeup chemical in the kraft mill.

V. CONCLUSION

The extensive interrelationship of the various process plant streams shown in Fig. 3 probably does not exhaust the possibilities. It is sufficient to show, however, that there may be more to be gained than conservation of energy alone by considering an integrated wood-products power co-generating facility. These include specific operational cost and capital requirements savings, and the possible generation of revenues from energy (electricity) production. Relative sizing of the various process units of Fig. 3, require a detailed, quantitative analysis of each part of the overall complex. Some of the required parameters can only be roughly approximated at the present time.

REFERENCES

1. Lars Elmenius, "By-Product Power - One Way to Cut Production Costs," TAPPI 57(7):76 (July 1974).
2. W. B. Wilson, "The Role of Turbines in the Paper Industry: How to Conserve Energy and Increase Profits," TAPPI 59(6): 95 (June 1976).
3. E. Gyftopoulos, J. B. Dunlay, and S. E. Nydick, "A Study of Improved Fuel Effectiveness in the Iron and Steel and Paper and Pulp Industries," Report PB-254042, U. S. Dept. Commerce, prepared for National Science Foundation by Thermo Electron Corp., March 1976.
4. Peter G. Belluschi, "The Total Tree Concept and Its Effect on the Pulp Industry," TAPPI 57(3):96 (March 1974).
5. Vance. C. Lischer, Jr., "Solid Waste as Supplementary Boiler Fuel for Paper Mills," TAPPI 59(6):104 (June 1976).
6. V. L. Hammond, L. K. Mudge, C. H. Allen, and C. F. Schiefelbein, "Energy From Solid Waste by Pyrolysis-Incineration," Paper presented by Battelle Northwest at the APCA meeting, Eugene, Oregon, November 1972.
7. T. F. Fisher, M. L. Kasbohm, and J. R. Rivero, "Waste Treatment Advances: The Purox System," CEP 72(10):75 (October 1976).
8. Lloyd H. Furman and Leland G. Desman, "Wood Residue for Veneer Drying - a Case History," For. Prod. Jour. 26(9):52 (September 1976).
9. John B. Grantham, "Status of Timber Utilization on the Pacific Coast," USDA Forest Service General Technical Report PNW-29, 1974, Pac. N.W. Forest and Range Exp. Sta., USDA, Portland, Oregon.

FEASIBILITY OF UTILIZING CROP, FORESTRY, AND MANURE RESIDUES TO PRODUCE ENERGY

J. A. Alich, Jr., F. A. Schooley,
R. K. Ernest, K. A. Miller, B. M. Louks
T. C. Veblen, J. G. Witwer, R. H. Hamilton

Stanford Research Institute
Menlo Park, California

I. INTRODUCTION

The realization that low-cost energy is no longer available has caused government and industry to place increasing emphasis on developing alternative sources of clean energy. Among these alternative sources, renewable energy resources such as fuels from biomass have received enthusiastic public support and increased government research, development, and demonstration funding. This paper presents the results of a U.S. Energy Research and Development Administration-Fuels from Biomass Energy Branch sponsored study* of the feasibility of using agricultural residues to produce energy.

* An Evaluation of the Use of Agricultural Residues As An Energy Feedstock: An Extension of Work, ERDA Contract No. E(04-3)-115.

It is estimated that these residues amount to 430 million dry tons within the conterminous United States. Of this amount, crop wastes constitute 280 million tons, livestock and poultry manures 30 million tons, and logging residues about 120 million tons.

Using a county-by-county inventory of the quantity of agricultural residues produced* including their availability, ten study areas of 50 miles in radius or more were chosen for feasibility analysis. For each of the sites chosen, the analysis included determination of:

(1) Residue characteristics, including the value or cost of current disposition, requirements for residue collection and transportation, and feedstock energy content.

(2) Residue condition, seasonality of production, location, and current disposition.

(3) The type of equipment required to collect and transport residues.

(4) Total energy production and consumption in the area.

(5) Specific product processing plant energy needs in the area.

(6) The potential feasibility of utilization of the residues for energy production.

(7) Institution considerations.

In this paper, the results of one site evaluation (Sutter, California), a summary discussion of other site studies and conversion projects that appear feasible, and general study conclusions are presented.

---

* Developed by SRI during the study entitled "An Evaluation of the Use of Agricultural Residues As An Energy Feedstock," for NSF/AERT, under grant No. 18615.

II. SUTTER, CALIFORNIA EVALUATION

The results of SRI's analysis of the potential feasibility of utilizing agricultural and forestry residues within a 50-mile radius of Sutter County, California, for production of energy are presented in the following paragraphs. The study area includes the northern California counties of Sutter, Butte, Colusa, Napa, Solano, Yolo, Yuba, and Sacramento, which cover an area of 7,679 square miles and supports a population of 1,271,000.

Agricultural and forestry residues in the area represent a substantial potential source of useful energy. However, while the major fraction of generated residues is available and collectible, in many cases the costs are high. Table I summarizes the estimated available quantities of residues and their costs for the Sutter area.

TABLE I. Quantities of Available Agricultural and Forestry Residues in the Sutter, California Area and Their Costs (Thousands of Tons per Year)

	As-received	Dry	Dollars/ dry ton
Dry mill residues	300	285	$ 1.00-110.00
High moisture field residues	675	310	23.85- 56.25
Low moisture field residues	1950	1576	13.75- 33.50
Vegetable field residues	845	170	nc[a]
Orchard prunings	235	83	27.15- 57.15
Forestry residues	600	340	2.85- 41.00
Animal manures	nc	70	nc
Total	4605	2834	

a. Not calculated.

The available quantity includes the residues that are collected during normal harvesting or judged to be realistically collectible. The delivered cost provides all costs through delivery to an energy conversion locality 15 miles from roadside or mill site. Costs for vegetable field residues and animal manures were not calculated because availability of these residues is scattered and relatively insignificant.

Of the residues in Table I, the most feasible for energy production on the basis of available quantities and costs are low moisture field residues. Table II shows a breakdown of the quantities of these residues by type and the costs of collection and delivery to an energy conversion facility. Forestry residues were excluded from consideration at this site, since forestry residues were addressed in detail at other sites.

TABLE II. Quantities and Costs of Available Low Moisture Field Residues in the Sutter, California Area by Type (Thousands of Tons per Year)

	As-received	Dry	Delivered cost ($/million Btu)
Barley	420	380	$1.05-1.85
Oats	45	39	1.05-1.85
Rice straw (baled)	905	725	1.50-2.15
Rice straw (chopped)			0.90-1.80
Safflower	80	72	1.05-1.80
Wheat	500	360	1.25-2.25
Total	1950	1576	

If the 1.6 million dry tons of low moisture field residues available annually were converted to energy, they could provide 2050 million kWh per year of electricity (1880 million kWh per year of electricity are consumed in the Sutter, California, area) or 13,300 million scf of synthetic natural gas per year (10,330 million scf per year are consumed in the Sutter, California, area).

Preliminary economics for a number of processes for the conversion of agricultural residues to energy were calculated, based on the residues and conversion processes shown in Table III.

TABLE III. Resource/Product Options

Residue type	*Conversion technology*	*Energy product*
Rice, wheat, and barley straw	*Gasification*	*Intermediate-Btu gas*
Rice, wheat, and barley straw	*Gasification*	*SNG*
Rice, wheat, and barley straw	*Combustion*	*Electric power*
Rice, wheat, or barley straw	*Combustion*	*Process steam*
Rice, wheat, or barley straw	*Combustion*	*Process heat*

Selection of the conversion processes was based on residue price and type and energy consumption in the area. The economics of electricity and fuel gas production were based on large scale utility facilities, while those for grain drying and process steam generation were based on industrial facilities.

The economics of combustion of rice, barley, or wheat straw in a 210 MWe central station steam electric plant are shown below as a sample of the economic evaluation.

	Mills/kWh
Processing cost	18
Residue cost	11-22
Total revenue required	29-40

The lower range of the net required selling price for electricity (∿ 30 mills per kWh) is competitive with alternative new steam electric options for the area. Analysis of co-firing the residues in a large coal-fired steam electric plant would result in even more favorable economics. This latter option is the most promising, since it also results in a reduction of the risk associated with fluctuating residue price and availability.

Production of SNG from residues also appears feasible, since the quantity produced would be relatively small and could be combined with existing utility supplies at a very low total cost increase to the consumer. A cost increase to PG&E's* customers of only 3% to 5% would be required if the expensive SNG from gasification of residues is rolled into PG&E's total gas supply. These percentages are based on PG&E's cost of gas in the year 1975. The percentages would be lower in future years at higher PG&E costs of natural gas.

Use of residues by industry in the area for direct heat or for process steam generation does not appear feasible for two reasons:

(1) High capital investment and low operating rates; and

(2) Lack of economics of scale.

Production of energy from residues will be influenced by a number of institutional factors. Important institutional considerations in the agricultural sector include concern that rice stem rot be controlled if rice straw is not burned in the field,

---

* PG&E - Pacific Gas and Electric Company, the large utility servicing many areas in Northern California.

regulatory pressure to cease field burning of rice straw, the capital expenditures required for collection and conversion equipment, and the fact that energy from regulated utilities is sold and purchased at average cost while energy produced by nonregulated industry is paid for at marginal cost. Regulatory decisions, system reliability requirements, and assurance of a long-term residue supply are the most important institutional factors to be considered by regulated utilities when considering conversion of residues to useful energy.

IV. SUMMARY OF CONVERSION CONCLUSIONS BY AREA

The following paragraphs list some of the projects that are feasible in each of the other study areas surveyed. A more extensive discussion of each of the conversion conclusions described can be found in SRI's Report entitled "An Evaluation of the Use of Agricultural Residues as an Energy Feedstock in Ten U.S. Areas," Volumes I and II, dated June, 1976.

A. Hendry, Florida

The conversion of blackstrap molasses to ethyl alcohol currently is technically feasible, and appears to be trending toward economic feasibility as the price of molasses stabilizes and the price of industrial alcohol from ethylene rises.

The conversion of bagasse to process steam or electricity at the sugar mill sites appears both economically and technically feasible at the present time. The economics of steam generation during the seven months in which the mills are shut down each year appears particularly advantageous. If the power potentially available can be put to use, idle sugar mill equipment can be utilized at a relatively low cost. This is discussed in the paper by Mr. Arlington.

B. Humboldt County, California

The conversion of mill bark and other mill residuals into process steam and electricity currently is feasible and would be particularly advantageous in the Redding area to solve a problem of the disposal of mill wood. In the most advantageous conversion a lumber mill would generate power to supply internal requirements, needs of nearby paper mills and/or local community requirements.

C. Sussex, Delaware

The conversion of poultry manure litter and wheat/barley straws to SNG through anaerobic digestion is feasible, provided the cost of feedstock is recovered through sales of sludge for fertilizer. The cost of generating electric power from soybean or corn stalk residues is about 30% higher than power currently purchased from the local utility.

D. State of Maine

Several feasible alternatives for conversion include the generation of process steam and electricity from millwood and bark. The low cost of fuel and relatively high cost of energy makes construction of conversion plants particularly advantageous in Maine. The utilization of wood has already been considered in several papers.

E. Weld, Colorado

The conversion of cattle manure to intermediate-Btu gas or SNG appears practical, especially if the feedstock cost is offset with a credit for the use of sludge as fertilizer. The low cost of cattle manure available at large feedlots provides a particularly advantageous source of feedstock material. This topic is further explored in the paper by Mr. Varani.

F. Traill, North Dakota

The high values placed on crop residues returned to the soil and the availability of large quantities of low-cost coal prevents conversion of these residues to energy from being considered economically feasible.

G. Bolivar, Mississippi

Electric power generation or process steam generation with cotton gin trash and soybean residues is both technically feasible and economically competitive with fossil fuel feedstocks and particularly advantageous at higher levels of utilization. The utilization of cotton gin residues is considered further, in the paper by Dr. William Lalor.

H. Lubbock, Texas

The use of cotton gin trash, particularly for conventional electric and steam power production is advantageous in the Lubbock, Texas area. Because of an inverted price structure (the mid-volume user pays higher costs per unit of electricity), the cost of electric power generation from cotton gin trash is about 87% of the small user's cost of electricity purchased for irrigation, and only 60% of mid-volume cotton gin operator's cost. Electric power produced from residues can be generated at a cost less than that paid by the small user. Therefore, the construction of residue fed electric power generating plants in this area appears feasible.

I. Piatt, Illinois

The conversion of soybean and field corn to useful energy products is feasible technically but not economically because of the large quantities of relatively low-cost high sulfur coal in this area. Costs of process steam and electrical generation using agricultural residues currently are about 20% to 40% more expensive than generation that uses coal as a feedstock.

J. General Conclusions

One of the most important factors in determining the feasibility of residue conversion to energy in a particular area is the price and availability of residues. Significant quantities of relatively low-cost residues were found in eight of the ten study areas.

Although 430 million tons of residue are theoretically available, representing enough feedstock to fire 90 electric power plants of 1000 MW capacity (80% load factor), or to satisfy roughly 20% of the nation's current demands for natural gas, in actuality residues would have to be diverted from other uses of higher value to achieve this level of utilization. Within the ten study areas, cash, feed, or fuel uses already have been found for about 20% of the residues produced; about 58% is returned to the soil, and, as shown in Table IV, 22% can be categorized as excess.

The definitions for the table utilization categories (Fed, Fuel, Returned, etc.) are shown in the Appendix.

In general, mill wood and mill bark are the most favorable residues for energy conversion because of the large quantities available, their relatively high-Btu content, the prices at which they are available ($0.20 to $0.65 per million Btu), and their location near energy consuming facilities.

The most economically feasible conversion possibilities exist with currently available and proven technologies, including process steam production, electric power generation, and anaerobic digestion of residues to produce methane gas. Frequently residue can be converted to energy using these techniques at costs that are competitive with the costs of energy produced at new facilities fueled with oil, propane, coal, or natural gas.

TABLE IV. Residue Uses: Ten-Site Survey (84 Counties)
(Thousands of Dry Tons Per Year)

Location	Fed	Fuel	Returned to soil	Sold or other	Not used	Total
Sutter County, CA	152		735	248	1,699	2,834
Humbolt County, CA		493		1,542	3,095	5,130
Piatt, IL	713		6,253			6,966
Traill County, ND	139		5,819	333		6,291
Sussex, DE			1,746			1 746
Maine		615		463	2,168	3,246
Weld, CA	311		558	375		1,244
Lubbock, TX	242		2,735		50	3,027
Bolivar, MS	59	20	2,096	183	191	2,549
Hendry, FL	4	786	426	230	270	1,716
Total	1,620	1,914	20,368	3,374	7,473	34,749
Percent of Total	5%	5%	58%	10%	22%	100%

APPENDIX

RESIDUE UTILIZATION DEFINITIONS

The categories shown in Table IV are defined as follows:

Sold - That portion of available residues that is collected and sold for any purpose other than for fuel in the case of forestry residues. All forestry residues used as fuel are included as fuel regardless of source or previous sale.

Fed - That portion of available residues that is fed to animals without collection, or with collection but without sale.

Fuel - That portion of available residues that is used as fuel without sale. This category includes all forestry residues used as fuel.

Returned - That portion of the available residues from crops and manure that is returned to the soil without sale.

Not used - That portion of the available residues that is disposed of at a cost (i.e., hauled away or burned). This category includes all logging residues in addition to collected residues that are not sold, fed, or used as fuel (i.e., they are returned to the soil at a handling cost).

LOGISTICS OF ENERGY RESOURCES AND RESIDUES

Thomas R. Miles

Beaverton, Oregon

I. INTRODUCTION

The goal of energy resource utilization is a processing complex, or complexes, that can produce those forms of energy and products yielding the maximum recovery from the various optimum combinations of available materials using known processes, and making best use of those proportions indigenous to an area.

Whether megawatts or methanol, pounds of steam or furfural, ammonia or hot raw gas, one should produce that which is either most needed at any price or that which is economically practical.

Those quantities and/or characteristics of the locally available agricultural and silvicultural raw materials can complement the ever present urban solid wastes to offer a reliable and continuing source of energy, albeit modest, but renewable and reliable.

Fundamental to the feasibility of any process are the logistics of raw material supply. The purpose of this paper is to attempt to put into practical perspective the various operational and economic factors and restraints of collection, storage, and transport that often dictate the feasibility of using renewable resources and residues as reliable energy sources. Other residue materials in addition to biomass are included.

Many resource investigations do not include these logistic considerations, probably because the investigator is preoccupied with the end use as a chemist, agronomist or combustion engineer or has no experience with the time, site requirements, manpower, and costs involved in the routine of supply. In like manner, this paper is only concerned with what happens beyond the conversion plant's door to the degree that the plant's desired form, density, bulk or unit package, and rate of demand affects the supply system's flow and economics.

Most of us here are involved with materials which might qualify as "soft technology," at least according to Amory Lovins. I would agree with Lovins to the extent that any renewable resource or residue is worth continued scrutiny. The concept of local or on-site use is very sound and is customary in the forest products, sugar cane, and pulp and paper industries as examples where residues are generated in the prime product processing plant. It should be used wherever feasible. Even at a lower utilization efficiency, local use of fuels and residues can often be far more thrifty of energy than the collecting-converting-redistributing process using central plants. This greatly simplifies the logistics problem.

Early locomotives and steamboats boarded fuel at relatively frequent intervals while combines and threshing equipment were fired by straw from the field being harvested. Homes were heated and food cooked in my own childhood home from wood cut within a few miles of home. Sawdust furnaces were common in homes in the Northwest until replaced in most homes about thirty years ago by more convenient oil, gas, or electricity.

Many people of the world still live immediately "off the land" for food as well as fuel, with a minimum of logistical problems--primarily because their needs are minimal--having gradually evolved with the land over millenia. This situation is changing, abruptly for some, both primitives and sophisticates. Our vaunted Western standard of living is characterized by the ready availability and convenience of virtually everything we use--food, clothing, shelter, transport, and entertainment--being energy intensive.

In the early days of my 30 years in engineering practice, I was busily removing steam plants from sawmills and replacing them with "clean" oil, gas, and grid electricial power; now I am putting them back. In the past two years I have been retained by a Committee of the State of Oregon to engage in intensive research and development of viable uses for the 500,000 tons of straw residual annually of the grass seed crops, both annual and perennial varieties, produced in Oregon's Willamette Valley. As one of the projects of resource recovery, we are just completing the design of a new straw-fiber production plant. One of the key processes in the plant's system is the drying of the 50% moisture (w/w) refined fiber. In the pilot plant we successfully tested flash drying the fiber using stack gases from a low emission double-vortex furnace developed for another purpose, fired with straw. Although straw is inconvenient for most users to handle, since the production plant is already using straw as a prime raw material, why not also as a fuel? It costs less than gas or oil and is in abundant supply. We cannot yet afford to produce our own electrical power, but that day may come. Parenthetically, however, it is probably because we are intent on straw utilization that we can put up with the inconvenience of handling straw.

To properly serve both producer and end-user, one must treat the whole sequence as a year-round system whose elements will var from material to material, region to region, and season to season. Further, one must remember that today's residue is tomorrow's resource, with consequent legitimacy as a material and with a recognized value. During the ensuing 10 to 15 years, we cannot implement as many complete or ideal system as are desirable. Established practices, continued relatively reasonable costs and availability of fossil fuels to extend the use of depreciated existing capital facilities, and time to establish the test demonstration units will all delay full realization.

Because my concentration has been largely in wood products and its waste utilization, and most recently in straw and cotton utilization, this paper will reflect these experiences. In general, wood waste utilization problems and their solutions are more clearly understood. In many cases these wastes have already worked back into the energy-producing systems of the mill. Straws are perhaps the most widespread renewable residues whose practical uses and logistics are only now becoming known.

II. FACTORS AFFECTING SUPPLY LOGISTICS

The majority of the following factors apply equally to both residues and resources. That particular combination of factors that results in a workable supply system for any given material must be economic in net energy as well as dollars.

A. Year-Round Sources

Year-round sources greatly simplify the establishing of stable personnel, full equipment utilization, bare minimum of storage needed and, if it is a point source, conversion of substantial quantities of material on the spot is often feasible, all contributing to a reliable supply of economic energy.

B. Seasonal Sources

Seasonal sources pose a host of special circumstances with storage an absolute necessity to provide a 12-month supply to any conversion plant. Single purpose harvest, collection or baling equipment is often required. Short removal seasons due to weather, cropping practice, or the particular cultivar characterize many crop residues such as cereal and grass seed straws, sorghums, etc. Many proposed biomass resources are seasonal also. Seasonal materials often require immediate drying, grinding, or other processing for preservation during storage as well as end use needs. Costs of capital tied up in stored material and processing must not be ignored.

C. Regional Distinctions

Generally the warmer latitudes provide longer seasons, with double cropping, or longer removal periods and/or multi-cutting of high yield biomass sources and residues. Northern latitudes or mountains can produce respectable forest biomass yields as well as substantial straw residues. Plains areas have a fairly consistent 2 to 3 tons/acre of available residues from normal crops. Urban residues are automatically near large energy needs.

A given region such as the Willamette Valley of Oregon may combine the stability of year-round sources such as industrial wood residues and urban municipal solid waste with the lower moisture and higher net energy of seasonal sources such as agricultural residues and forest slash. Experimentation has shown that both municipal refuse derived fuel (30% MC) and wood residues (50% MC) are highly miscible with coarsely shredded grass straw (10% MC) when compacted or extruded for fuel processes.

D. Weather

Weather can badly disrupt harvest schedules, especially with crop residues like straws and corn stalks which can only be harvested after the prime crop is in. On occasion this delay can nearly destroy the residue by forcing harvest into fall wet weather resulting in a wet,moldy product. Additionally, such things as floods or drought can virtually wipe out the yield and recovery in some areas. Extremes of cold or heat can sometimes be an advantage such as harvesting freeze-dried corn stalks in midwinter on hard frozen ground. Midday or very low humidities create cereal or grass straw baling problems with either round or standard bales. The stiffness of the stems causes pick-up problems and low bale densities.

E. Field Conditions

Field conditions can be particularly difficult, and costly, with crops like rise straw with its small diked muddy paddies. Equipment which compacts or ruts the fields must be avoided especially on perennial crops. Steepness of slopes affects equipment requirements and rate of collection.

III. RESIDUE VALUE

The prime crop bears the "swathing" and combining costs and can leave the residue in a windrow. Residues must bear their own collection costs and manpower-management arrangements. There are various opinions and practices regarding the "in-the-field-value" of these residues to the farmer or owner. Corn husklage has been valued at $8.00/GT (green ton). Straws range from gratis to $10.00/T in the field. It does make a considerable difference in improving the grower-user relationship if some reasonable value is placed on the residue, preferably in a long-term contract.

With crop residues, seldom does the farmer bale or otherwise remove his own residues, but depends on contract in a collection and transport. These services constitute the primary costs of the residue. Farmers have had some field damage and harvest delay problems with residue contractors and are thus reluctant to depend on untried contractors, but they would still prefer to contract these activities to a reputable firm.

IV. COLLECTION

Collection or harvest methods are critical to the success of either residues or resources utilization. In some instances, the residue is carried into the conversion plant along with the prime product. Sawmill and veneer plant hog fuel and sugar cane bagasse are examples. However, even with these materials, there are additional leavings in the woods or fields that can be recovered when economically feasible. Nearly all agricultural residues must be separately collected and cost accordingly.

Ideally, collection equipment, or part of it, can serve usefully for other duties but it is not uncommon for depreciation and maintenance alone to cost to $3.00 to $5.00/ton for specialized harvesting of single season crops or residues.

As the first step in any supply system, the collection method must be considered in terms of its effect on the succeeding operations. Although last in line, the needs of the conversion plant as to the desired form, shape, density, load type and size of the material will often determine not only the collection method but storage and transport methods as well.

With the exception of the baling methods used for straws and grasses, nearly all collection systems are essentially bulk handling from the field or hillside. Roadside bundling, module building, chipping, grinding and/or compacting operations often receive these bulk materials from the field preparatory to transport and/or storage. Stackwagons with flail pick-up making 8' × 20' × 9' stacks of straws, reeds, or stalks are very versatile and the most economic collection system, as well as producing a self-storing stack, when properly thatched over. Allowing 3 ft aisles, 172 stacks can be stored per acre or a total of 430 T/A (with 5000 lb straw stacks) can be stored. This would cost approximately $1.00/T/yr for land use. Drainage and accessibility are most important. This system works well for short hauls (3 to 6 miles) to the conversion plant or especially for on-site use. Roadsided costs range from $8.00 to $12.00/T.

Round bales have become very popular in the U.S., and particularly in the last two seasons in the United Kingdom and Denmark. Our experience in Oregon indicates that they are fine for on-farm use; they do not weather well in damp climates (as the UK has also discovered). They cannot be economically stored under cover or hauled any distance, at least when made from stiff, light straw stalks. The net bulk density of round bales of straw on a truck is approximately 2.5 lb/C.F. and the balers are encountering operational and bale handling safety problems.

The collection of "plantation" resources can be more efficient than seasonal residues since there is no "prime crop" to delay or complicate an orderly operation. High removal rates of residues are often necessary to permit cultivation or preparation for the next cutting or crop.

Kelp, hyacinths, and algae all have their own collection conditions and specialized equipment which operates year-round on these resources. They are all perishable by nature and must be quickly (1 to 3 days) processed.

Logging of natural forests or plantations is a well established business with efficient equipment. Logging residues (slash) are being given a good look by those ingenious loggers and equipment suppliers who have so far met all challenges in timber handling and supply. Two general classes of residues are left, chippable larger pieces and miscellaneous limbs, branches, and broken pieces which probably should be hammer-milled and air-classified to remove rocks and soil. A study of a series of logging sites in Oregon and Washington by Dell and Ward showed an average of 89 tons of residue per acre of which 41 tons was chippable. Estimated collection costs by Grantham and others were from $14.00 to $17.00/BDT and transport $12.00-$15.00/BDT for a one-way 70 miles haul in 1974. Field processing could possibly reduce the hauling cost by providing full loads.

Mesquite (10 to 15 T/A) in Texas for fuel, jojoba in Arizona and New Mexico for sperm oil replacement and guayule for rubber will each obviously have their own systems of collection as resources.

V. WHOLE HARVEST

Whole harvest of the crop and its residue, bringing them into a central processing plant to be dried, threshed, chopped, ground, and fractionated as necessary for best use is receiving a great deal of worldwide attention. We experimented with a partial system in 1974 in Oregon on grass seed and straw. Kockums is trying this season (1977) a very elaborate system in Sweden to do 2000 acres of barley and wheat. The University of California at Davis is about to launch a whole harvest program for rice in the Sacramento area and undoubtedly there are others. It will make sense:

(a) If a ready market or use for the residue exists, and

(b) If the high capital investment in processing will pay out with seasonal use.

Processing and threshing are necessarily immediate with whole harvested crops, which shifts the storage problem to the final products and their storage must be enclosed. In most of

TABLE I. Crop Residue Collection Systems and Costs

Collection method	Material	Moisture (%)	Field density T/A	Collection rate T/hr	Unit size & wt	Density lbs/CF	Approx. roadside cost/T ($)
From windrow							
Standard 3-wire bale	Hay	15-18	5-8	8-15	16×22×46 110-130 lbs	10-14	12-16
	Cereal & grass straws	10-12	2-5	8-10	16×22×46 89-90 lbs	8-10	13-19
Hi-Density Bale	Cereal & grass straws	10-12	2-5	6-8	14×18×44 100-110 lbs	14-16	15-20
Stackwagon	Straw	10-12	2-5	7-12	8'×20'×9' 5000 lbs	2.5-4	8-12
Round bale, 7 ft	Straw	10-12	2-5	3-5	7'dia×6'lg 1200 lbs	5-6	12-16
Forage harvester[a] 1" reel chop into compress. trailers	Straw	10-12	2-5	5-6	Bulk	5-6 in tr'lr (2.5-4 loose)	14-19
Stackwagon	Cornstalks	50	2	n.a.	8'×14'×8' 6000 lbs	7-12	n.a.
Field cube[b]	Hay	15-18	5-8	2.5-4	1¼×1¼ cubes Bulk	18-20	16-25

a. Includes chopping and 5-mile haul.
b. Straws will not cube satisfactorily without a binding agent.

the grains and grasses, the collectible residues will range from 1.5 to 4 T/A. Areas with normally wet harvest seasons may realize overall economy by using this predictable whole harvest method.

TABLE II. Densities and Moistures of Other Residues

Materials	*Moisture (%)*	*Field density*	*Unit share*	*Density lb/CF*
Wood chips	*45-55*	-	*Bulk*	*20-24*
Wood hog fuel	*30-50*	-	*Bulk*	*15-17*
MSW	*25-35*	-	*Bulk*	*9-15*
Corn husklage	*50*	*1*	*Cotton module 7'×32'×8'*	*9*
Cottin gin trash	*10*	-	*Bulk*	*3-5*
Bamboo	*50*	*7*	*Bundles*	*8-10*

VI. STORAGE

Storage of residues and resources is a fact of life with most seasonally produced materials. Whenever possible, materials that require preprocessing such as drying should also be densified and unit packaged to more efficiently use the covered or enclosed storage that is obviously then required, as well as expediting subsequent removal and reducing transport costs.

A. Outside Storage

Outside storage can be used in some climates with certain materials if properly stacked or piled. Most green wood residues, baled or chopped straws with carefully thatched straw covers, and bagasse are examples. Wet area residues can be moved toward end use points directly from the field to storage in dry climate areas. Thus the need for enclosed storage can be eliminated. The example in Oregon is straw removal from the valley to storage in Central or Eastern Oregon. Storage sites should be well drained, with surveillance to deter vandalism, and with storage piles not prominently evident. Drainage of polluting leachates may occur and must be contained. Outside storage costs include cost of land as well as labor to put up stacks, or unload, and can range from $1.00-$3.00/ton for bulky materials. In some instances the collection or harvest equipment can "roadside" the material in units or modules directly from the field to all-weather accessible areas for later transport to conversion plant.

One should avoid stacking materials containing simple carbohydrates or sugars when too damp (over 18% w/w) to prevent spontaneous combustion. Also damp conditions encourage mycotoxin-producing molds (viz. A. flavious whose toxins are such that the material cannot be used for feeds). Mild NaOH treatment will usually neutralize these effects, and inhibit mold growth (over a pH of about 9). Propionic acid is also in use as a preservative for bagasse, hay, and straw. Plastic stack coverings have proven very unsatisfactory.

B. Enclosed Storage

Enclosed storage enecompasses many types and degrees of buildings with protection from rodents, micro-organisms and people as well as weather. Systems range from single roofs to silos. Storage sheds serve also to protect equipment.

A simple roof, 16 ft in the clear, with two sidewalls and a rocked floor will store 9-10 lbs/CF material for $5.00/ton/yr or 16 lbs/CF material for $3.00/ton/yr. Completely closed, concrete floored storage will double these costs at the same densities, but will store highly densified bulk material such as 1/4 in. pellets (40 lbs/CF) for $2.50/ton/yr.

Whenever possible, on-farm, plantation or site storage of seasonal residues or resources allows efficient year around use of transport equipment and avoids the scramble for both rigs and drivers that occurs every harvest season in any growing area. The number and size of needed collection and transport equipment units and their costs should be estimated for a given residue and end use. It is too often assumed that a farmer has balers, or tractors, or trucks etc. sufficient to the task, or that such equipment is readily available to buy or lease. Most collection systems either can, or should, deliver the seasonal materials to a roadside storage area or building. This avoids congestion at a conversion plant, disperses storage risks and develops a steady cash flow.

C. Dust Explosions

Dust explosions caused by processing or storage of very dry residues can be prevented by keeping the material moisture content over 10-12%. Our tests indicate that cellulosic residue dusts (except plywood sander dust) tend to burn rather than explode if amply vented and kept over 10% moisture. As a precaution it is wise to use building designs with no beam ledges or horizontal girt surfaces to collect dusts. Heavy timbers or glued laminated beams will withstand flame much longer without collapsing than steel. Enclosed spaces should also have sprinklers for permanent protection.

It is also well to remember that many stems and tubular materials such as straws contain sufficient air to sustain limited combustion, making it virtually impossible to completely quench a fire in them. Separated stacks of 200 tons each, separated by 60 ft wide "fire lanes" as well as 20-30 ft cross access lanes for loaders and stackers, and well-distributed fire hydrants with hoses, nozzles and a good water supply are all highly recommended. Whenever possible, the long dimension of the stacks and the yard should be oriented across the prevailing wind to limit fire spread.

D. Rodent Control

Rodent control is important where residues will also be used for feed or fiber products. Rat feces and urine destroy palatability of straw and carry diseases which may be fatal to horses and cattle. Closed buildings should not be accessible from the outside. They should be well drained, and straw residues kept separate from other feeds such as grains, manures, or liquids. Rats nest comfortably in NaOH treated straw. Barn owls control mice and moles but not rats. Cats control mice but also bring field mice into storage areas. They do not control rats.

A recent infestation of 50 rats in 200 tons of straw in an enclosed building was eliminated by:

(1) Effective cleaning and removal of potential habitats,

(2) Rat bait placed in key areas, and

(3) Daily removal of bodies and rebaiting.

Control took 5 weeks, 50 pounds of rat bait at $1.00/lb or a cost of $1.00/rat.

VII. TRANSPORT

Transport is a critical and limiting factor, both in cost and availability. The form and condition of the material controls loading and unloading time, and whether open or closed trucks are used. The density controls the payload, as illustrated in Table III. Year-round sources can be organized on a regular basis and be predictably reliable and economic. Table III can serve as a density guide for most general freight movements and contract carriers.

TABLE III. General Transport Volumes and Densities

Size	*Volume (cu ft)*	*Payload (lbs)*	*Minimum bulk density for payload (lb/cu ft)*
Truck-trailer			
(2) 27 ft flat bed	*2,000*	*55,000*	*14*
(2) 27 ft dry box	*1,771*	*55,000*	*16*
40 ft flat bed	*3,200*	*52,000*	*16*
40 ft dry box	*2,650*	*50,000*	*19*
Rail			
50 ft	*4,960*	*100,000*	*20*
40 ft	*3,750*	*100,000*	*27*
Container (Offshore)			
40 ft	*2,000*	*44,000*	*22*
20 ft	*1,050*	*40,000*	*38*

A. Costs of Transport

Costs of transport will vary considerably depending on four factors: load density, regularity, load and unload conditions, and distance.

Generally, chopping or grinding a stem or leafy material will increase its density, but only to a point which varies with material. Some load-compressing trailers or modules can improve densities up to about 6-8 lbs/CF for dry materials.

Three different trucking options can be considered: common carrier, contract hauling, or company-owned-and-operated equipment. Common carriers charge general cargo tariffs which are quite high in comparison with contract haulers who serve best on either year-round or long-season point-to-point or area-to-point runs. Erratic demands and/or specialized equipment requirements may result in the necessity for a conversion company to operate its own equipment.

Contract carrier charges provide the most reliable cost information. Typical charges in the Pacific Northwest for point-to-point daily service hauling wood residues with one-half hour load and one-half hour unload times are $1.50/one way mile plus $25/load under 30 miles and $20/load over 30 miles with no back haul.

Table IV illustrates the costs of various densities and distances. The selection of these two densities is based on a minimum bulk density of 16 lbs/CF for a full payload (Table I) and 10 lbs/CF as the commonly realized maximum density of many agricultural residues including 3-wire bales of straw (Table II). This data is for an over-the-highway, first class truck-trailer, either flat bed or doubles and includes all costs of ownership and operation of truck and equipment and based on year-round or long season operation, for contract carrier rates in the Pacific Northwest at the present time (July 1977).

The tabulation in Table IV illustrates the fallacy of using a ton-mile basis for estimating transport costs. It is best to work each supply situation and material out as it develops. Also there is the exasperating, but nevertheless real, difference between the states (and sometimes counties) in truck regulations. Other transport problems to be considered are the lack of ICC load classifications for many agricultural residues, the PUC regulations regarding farm-owned and operated vs. contract carrier equipment, and in some cases union jurisdictional disputes.

TABLE IV. Contract Truck Costs, Load Density and Distance

	10 miles		20 miles		50 miles	
	10	16	10	16	10	16
	(pounds / cubic feet)					
Truck-trailer load (tons)	17	27	17	27	17	27
Mileage charge at $1.50 ($)	15	15	30	30	75	75
Fixed charge/load ($)	25	25	25	25	20	20
Total Cost ($)	40	40	55	55	95	95
Cost per ton ($)	2.35	1.48	3.23	2.04	5.59	3.52
Cost per ton-mile ($)	.23	.15	.162	.10	.11	.07

VIII. EXAMPLES OF LOGISTICS AND COSTS

The following examples will illustrate the various applied logistics and costs involved in supplying a conversion plant.

A. Year-Round Residues

1. *Sawmill and Veneer Residues*

Sawmill and veneer residues constitute a substantial part of the log, as illustrated by Table V. Note that wood chips are included as products rather than residues. Chippable wood was a residue--and a mill disposal problem--until about 1950 when it approached commodity status. In about 1960 it became the basic raw material for pulp and paper, largely replacing "round wood" or logs. Now worth $30-$40/BD ton at the mill, chips contributed $18-$24 for every thousand feet of logs sawn and accordingly influence timber prices.

TABLE V. Products and Residues From Sawlogs

	Products (%)	Residues (%)	Gr. wt/ cu. ft. (lbs)
Lumber	40		
Wood chips	25		22
Sawdust		12	18
Shavings		10	9
Bark		13	15
Total	65	35	
Mill-run blend of sawdust, shavings, bark			16

My firm was deeply involved in the development of sawmill chipping equipment, methods of production, etc. accomplishing more than 75 plant installations in the 1950s and 1960s. We also developed the currently prevalent techniques to "compact load" wood chips to a density of 22 lbs/CF gross weight, getting as much as 50% more in the rail car than previously. This of course immediately increased the economic shipping radius by 50%, and eventually to the extent that wood chips have been regularly hauled from Colorado and Wyoming to Wisconsin by rail for the past 10 years, some 1200 miles. Many hundreds of special chip cars and trucks are now in turn-around service nationwide, another evidence of the evolution of a residue to a resource.

The remaining sawdust, shavings, and bark can be converted into sufficient electrical and thermal energy to operate the mill, or if logging residues are included, a surplus of electrical energy can be sold to the grid along with some additional fiber and possibly chemicals.

There is currently a steady reconversion by sawmills to replace as many oil or gas uses with wood residues as possible.

Many are considering but few are actually installing generation equipment.

Current prices paid for sawmill "hog fuel" in Oregon are \$4-\$7/unit or BDT, delivered within 10-15 miles of source mills, while logging residues will have to be collected separately and hauled 30-70 miles and cost \$25-40/BDT. One firm in Oregon has started producing 1/4 in. diameter pellets made from hog fuel and selling the product for \$22/BDT or \$1.39/MM Btu.

2. Municipal Solid Waste

Municipal solid waste (MSW) is unique in that its total disposal is mandatory and the cost is paid by the source, the homeowner. Along with used tires and sewage sludge, it is urban oriented, but can be seriously considered as a reliable base energy supply to be used in combination with other locally available residues or resources, as attested by the many demonstration projects underway.

Owner pays \$4/month or \$40/ton		\$ 40.00
Collection and overhead	\$28.50/ton	
Haul to landfill	9.00/ton	
Landfill cost	2.50/ton	
Total cost	\$40.00/ton	40.00
Net cost at conversion plant		\$ -0-

The presegregation of waste paper of specific grades from MSW is a very healthy sign of conservation progress, but it is also economic to do so and is therefore being done.

3. *Year Around Resources*

Year around resources in the northern latitudes are somewhat limited in variety. Tree plantations, both conifers and hardwoods, apparently can best utilize the climate with minimum tending. The new "super trees" can do very well. Mr. Robert L. Jamison of Weyerhaeuser Timber Company presented a very good paper including logistics on this topic at IGT Symposium in Orlando 1977. He indicates a plant delivered cost of \$20-\$25/BDT for low grade hardwoods and presents a real need for dual fuel-fiber use in the next 10-15 years. Some kelp is harvested in northern waters year around.

Warmer sites can produce a host of continuous resources including trees and kelp, and also algae, hyacinth, various grasses, and bamboo.

B. Seasonal Residues

Seasonal residues include bagasse, corn stalks and husklage, cannery wastes, cotton gin trash, sorghums, reeds, rice straw, and cereal and grass straws. Park and garden trash is a particular nuisance to cities and homes and fills up landfills. In circumstances of severe drought or infestations, wasted agricultural debris could conceivably be processed for fuel, if properly situated near a plant, rather than destroyed in the field.

The following example of a grass straw supply system can be considered fairly typical of cereal straws also.

TABLE VI. Typical Conditions and Costs for Collection and Delivery of Oregon Grass Straw to Local and Long Haul (250 mile) Feed and Fuel Markets

Conditions	
Location	Willamette Valley, 300,000 acres
End use area	Local, Eastern and Southern Oregon (250 miles)
Seed harvest and straw collection season	July-September
Weather	Dry or intermittent rains
Average residue yield	2 tons/acre
Total residue available	500,000 tons removable
Suitable for feed	350,000 tons
Existing collection/sale	50,000 tons
Competing feed products/residues	Alfalfa hay, cereal grains and straws, cannery residues, and potato wastes
Competing solid fuel products/residues	Sawdust, hog fuel, MSW, logging slash

Costs	Unit $/ton	Accum. $/ton	Observations
Residue value	4.00		In the windrow
Collection	8.00		Standard bale 6 T/hr, 8 lb/CF
Roadside	3.00		
Approx. roadside cost		15.00	
Load, transport to storage	5.00		16T (2-27' flat bed w/5' extensions)
FOB storage		20.00	356 bales
Covered storage/yr	5.00		
Stored cost		25.00	12-mon. supply basis
Load and transport	5.00		
FOB 15 miles		30.00	E.G. fuel or fiber plant
OR	OR		
Load and transport	20.00		
FOB 250 miles		45.00	

The costs presented in Table VI are within \$1-\$2 of actual costs for several recent transactions (1977).

Seasonal resources include kenaf, various maizes, cattails, sudan grass, and many others being tested. Most can be collected, etc. by methods being developed or in use for other materials. The most significant differences between resources and residues are that the planting, row spacing, etc. are designed for whole harvest methods and that the total crop cost is now charged to the conversion use.

Table VII lists the various fuels available in Western Oregon and their costs. Materials and values will change from area to area and with time.

IX. CONCLUSIONS

The problems of collecting and delivering energy resources to a conversion facility are clearly reflected in present costs of wood, municipal and agricultural residues. Interestingly enough a number of renewable or recurring residues and resources appear to cost about the same to collect, store, and transport in the \$20-\$30/BDT range. It would be well to consider using several raw materials for a given facility or region.

Not only should the logistics of each residue or resource be individually evaluated but other nearby materials should be studied for:

(1) Substitution in case of shortage,

(2) Combination in processing for greater product or value yield,

TABLE VII. A Comparison of Approximate Raw Material Costs for Fuel or Fiber (Revised to June 1977, FOB Willamette Valley)

Material Gross Btu Value	Unit Cost ($)	Fuel Cost $/MM Btu
Oil, diesel 145,000 Btu/gal	0.39/gal	2.69
Oil, heavy $12.13/42-gal bbl	0.29/gal	2.00
Natural gas Therm = 100,000 Btu = 100 cu ft	0.20/therm	2.00
Coal, Wyoming 12,000 Btu/lb	50.00/ton	2.08
Hog fuel (mostly bark) Unit = 2000 lb bone dry 8000 Btu/lb	5.00/unit	0.31
Pelleted hog fuel 9000 Btu/lb	25.00/ton	1.39
Wood chips Unit = 2000 lb B.D., 8000 Btu/lb	40.00/unit	2.00
Straw, baled (12-mon. supply) 8000 Btu/lb	27.50/ton	1.72
Municipal refuse (MSW) 5000 Btu/lb	(5.00/ton)	(0.50)
MSW + straw 6500 Btu/lb	12.00/ton	0.92

(3) Similar or common equipment that can be used for staggered season collection, transport, or conversion to broaden the source base and increase equipment utilization. Changeable beds on trucks and heads on combines are examples, and

(4) Diversity of operation to accommodate changing markets or source.

From our experience the development of improved collection, stroage, and transport equipment and methods is going to be up to those of us in the field rather than the large manufacturers, at least until a volume market develops.

When utilizing a forage-type residue such as some straws, there may be times of drought, as now, when the feed requirements supersede any other use. Potential fuel resources must also be evaluated for alternative market opportunities, e.g., feed, fiber.

All of the foregoing logistics are predicated on available fossil fuel for motive power. We must eventually consider how to achieve self-sufficiency of cultivation and transport fuels, preferably locally from the residue. Finally, it could be that I have spanned the apogee of the trajectory of the Convenient Energy Curve--from splitting kindling to splitting kindling in 60 years.

REFERENCES

1. J. D. Dell, and F. R. Ward, "Logging Residues on Douglas-Fir Region Clearcuts--Weights and Volumes." USDA Forest Service Research Paper PNW-115, 1971.
2. J. B. Grantham, et al. "Energy and Raw Material Potentials of Wood Residue in the Pacific Coast States." A summary of a Preliminary Feasibility Investigation. USDA Forest Service General Technical Report PNW-18, 1974.
3. W. F. Buchele, "Harvesting and Utilization of Cornstalks from Iowa Farms." Report #3, Iowa State University Agricultural Engineering, 1975.
4. R. L. Jamison, "Trees as a Renewable Energy Resource." Symposium" Clean Fuels from Biomass and Wastes, sponsored by Institute of Gas Technology, Second Annual Symposium, 1977.

BAGASSE AS A RENEWABLE ENERGY SOURCE

William Arlington

Florida Sugar Cane League, Inc.
Clewiston, Florida

I. INTRODUCTION

During the past four years the price of fuel oil has quadrupled. Such an increase has had a significant impact on high energy consuming industries. In the production of raw sugar from sugar cane, energy demands are high. The domestic raw cane sugar processing industry uses approximately 50% as much energy, annually, as that produced for the City of Chicago. This amount of energy would require the expenditure of about 200 million gallons of fuel oil or its equivalent. However, the sugar cane industry has an advantage in that one of its major byproducts is utilized to produce energy.

Bagasse is a fibrous material obtained after the sugar cane passes through the final set of rollers in the sugar extraction process. The fibers vary in size according to the processing and cane variety. Bagasse is composed mainly of cellulose and uncombined water. Also included in its composition are xylan $(C_5H_8O_4)n$, lignin $(C_{24}H_{26}(CH_y)_2O_{10})$, cane wax, organic acids and other materials associated with plant life.

Since bagasse is an indefinite material, an exact carbon, oxygen, hydrogen content cannot be determined. However, a number of analyses show bagasse to contain about 5% to 7% hydrogen with the balance made up of approximately equal parts of carbon and oxygen on a dry basis.

II. THE USES OF BAGASSE

Bagasse has a number of uses. It can be pelletized for cattle feed, using molasses as a binder. It can be pressed into wallboard or used to produce furfural. It can be used to produce paper. This use is particularly significant in developing countries. However, its main use is to produce energy for the sugar mill [1].

III. ENERGY UTILIZATION OF BAGASSE

Most sugar mills are equipped with electrical generators to supply electricity to the mill. This provides economical electricity and helps to balance boiler steam demands. In some areas such as Hawaii, electricity is supplied to nearby towns as well as the sugar refinery. This co-generation concept could be used throughout renewable resources, as the paper by Dr. Johanson and Dr. Sarkanen demonstrates.

The bagasse is distributed to the boilers by a conveying system comprised of moving slats connected by chains on both sides. The bagasse drops through a trap door which controls the rate to the feeder. The feeder is a drum shaped housing with rotating blades that moves the the bagasse to the boiler at a uniform rate.

A. Types of Boilers

Two types of boilers widely used by the sugar industry are cell type furnaces and spreader stokers. The cell or horse shoe type furnace is comprised of from three to six cells. It is fired by distributing the bagasse in a pile to each of the separate cells. Vents are located along the outer walls on the bottom of each cell to allow for furnace draft. This type of furnace is simple in design. There are no moving parts to malfunction. However, obtaining maximum efficiency is a problem since combustion takes place toward the surface of the pile. Another problem of such furnaces is that the boiler must be cleaned by shutting down one cell at a time which causes the boiler to operate at less than maximum capacity.

The spreader stoker type furnace employs a traveling or vibrating grate. Combustion of the smaller particles actually takes place in suspension, in the air above the grate. The ash is then dropped below the grate into an ash pit.

Obtaining the proper fuel-air ratio is one of the greatest operating problems of a bagasse fired boiler. The major reason is that the moisture content of the bagasse is high, usually slightly above 50%. The moisture varies with the processing, making it difficult to stabilize operating conditions of the boiler. Both types of boilers use varying combinations of induced and forced draft to control air mixtures to the combustion chamber.

B. Boiler Operation

In any combustion process the theoretical amount of air can be calculated to determine the exact amount necessary to carry out the oxidation reaction. However, this amount is always exceeded in order to provide ample oxygen for the reaction to proceed. In the case of bagasse boilers, this excess air is also necessary to provide a media to absorb the steam created by the high moisture content of the fuel. Because of this, it is especially difficult to keep the amount of excess air at a minimum and thereby reduce the volume of cool air entering the furnace. Large amounts of cool air will cool the combustion area and the furnace itself, reducing efficiency. Many boilers have been equipped with air preheaters to reduce the cooling effect of incoming air. Hot exhaust gases are ducted in such a way as to heat the air before it enters the furnace. In this way some of the heat value of the exhaust gases is transferred back into the furnace rather than being lost out the stack.

Another method of increasing furnace efficiency is the use of bagasse dryers. Various types of dryers are used but the principle is the same. Hot exhaust gases are passed over the bagasse using a large container and controlling the contact time to ensure sufficient drying. These systems are not widely used for a number of reasons:

(1) Drying systems require the use of large amounts of energy.
(2) They increase the risk of fire or explosion.
(3) The dried bagasse is often swept through the furnace too fast to burn, increasing particulate emissions and decreasing furnace efficiency.
(4) The dryers themselves are considered as a source of particulate emission.

A major problem associated with the use of wet bagasse is that an alternate fuel is usually needed to aid combustion in larger boilers. Many of these boilers are equipped to burn oil when necessary. This is usually performed when energy requirements change or the supply of bagasse to the boiler is interrupted. However, under ideal conditions bagasse will supply the energy necessary for the milling operation.

C. Pollution Control

The final difficulty with the burning of bagasse as a fuel to be discussed here is that of pollution. Due to its relatively low heat value, higher amounts of fuel must be used. Because of this, greater amounts of particulate emissions are produced making it necessary in Florida to install scrubbers to reduce the amount of particulate emitted. Considerable cost is involved, not only for the purchase of these devices, but for maintenance as well. The type of scrubber most widely used in Florida is a wet scrubber. The use of such scrubbers eliminates about 90% of particulate emissions. Bagasse contains only trace amounts of sulfur and therefore sulfur oxide emissions are far below that of most other fuels.

IV. THE ENERGY CONTRIBUTION OF BAGASSE

The heat value of dry bagasse varies with cane variety, the fiber size, and external area. For the purpose of calculations, Deerr [2] used a figure of 8350 Btu/lb of dry bagasse. However, bagasse as used in the boiler, on the wet basis has a far lower heat value. Since this value varies greatly, it is impossible to select an exact figure. Again for the sake of calculation, a value of 3250 Btu/lb is used as the heat value of one pound of bagasse on the wet basis or about one-sixth that of fuel oil.

Energy demands for the production of raw sugar are high. In Florida it has been determined that a boiler, under normal conditions, imparts 1050 Btu to each pound of steam; and that an average of 950 pounds of steam is necessary to grind one ton of cane (or approximately one million Btu for each ton of cane ground). Using this figure as a base along with the total number of tons ground, the total energy requirement can be determined. According to Florida production figures for the 1976-77 crop year, just under ten million tons of cane were ground. Using the actual figures, 9.9×10^{12} Btu were consumed by the Florida sugar industry in processing cane into raw sugar. It is estimated that the domestic cane sugar industry will consume 3.17×10^{13} Btu during the 1976-77 crop year.

Such energy consumption by the domestic cane sugar industry without the use of bagasse would require the expenditure of 214 million gallons of fuel oil or its equivalent. Presently 66.1% of the energy consumed by the Florida sugar industry is generated using bagasse. Applying this percentage to the domestic sugar industry, an estimated fuel oil savings of 141 million gallons or its equivalent will result during the 1976-77 crop year from the burning of bagasse as a fuel. These calculations are summarized in Table I. The amount of bagasse burned to effect this savings is an estimated 6.45×10^{9} pounds or approximately 40% of the total amount produced. At a density of 12.5 lb/ft^3 such an amount would occupy 5.16 billion ft^3, creating a serious storage problem if it were held for other usage. The storage of large amounts of bagasse is a fire hazard. The removal of such amounts of bagasse from the milling area would be costly. Because of these factors, the use of bagasse as a fuel is the most efficient method of use and disposal.

TABLE I. *Energy Table (1976-77 Crop)*

Florida Sugar Industry	
Fuel oil burned	1.82×10^{8} lb
Heat value of fuel oil	18,500 Btu/lb
Total heat derived from fuel oil	3.36×10^{12}
Percent of total energy derived from fuel oil	33.9%
Bagasse burned (wet basis)	2.02×10^{9} lb
Heat value of bagasse (wet basis)	3,250 Btu/lb
Heat derived from bagasse	6.56×10^{12}
Percent of total energy derived from bagasse	66.1%
Percent of bagasse used for energy	40.7%
Approximate energy required per ton of cane	1×10^{6} Btu
Total energy utilized	9.92×10^{12}
Domestic Cane Sugar Industry [a]	
Bagasse burned (wet basis)	6.45×10^{9} lb
Heat derived from bagasse	2.09×10^{13} Btu
Total energy utilized	3.17×10^{13} Btu

a. Estimates extrapolated from Florida statistics.

IV. CONCLUSION

In conclusion, the use of bagasse as a fuel for the sugar industry is an efficient means of generating the energy necessary while conserving non-replenishable energy sources. Since it is likely that the cost of fuel oil will continue ro rise, there will no doubt be a continuing effort on the part of the cane sugar industry to conserve even greater amounts of fuel oil.

REFERENCES

1. Edward J. Lui, "The Hawaiian Sugar Industry Looks at Energy Farms." Typescript of presentation at the Conference on Fuels from Sugar Crops, Battelle Columbus Laboratories, Oct. 15, 1976:F-1, The Hawaiian Sugar Industry Looks at Sugar Energy Farms.
2. N. Deerr, _Cane Sugar_. Norman Rodgers, London, 1921.
3. A. C. Barnes, _The Sugar Cane_. John Wiley and Sons, Inc., 1974.
4. J. G. Davies, _The Principles of Cane Sugar Manufacture_. Norman Rodgers, London, 1938.
5. F. O. Lichts, _World Sugar Production, First Quarter Production 1976-77_, Ratzeburg, Germany, 1977.

USE OF GINNING WASTE AS AN ENERGY SOURCE

William F. Lalor

Cotton Incorporated
Raleigh, North Carolina

I. INTRODUCTION

A combustible waste is produced in the process of ginning cotton, a process in which heat is used for drying the crop. It was postulated that sufficient heat could be recovered to meet the drying needs if the waste were burned and about 30% of the heat were recovered. A two-season study conducted at gins, where heat-recovering waste incinerators were in use, confirmed the correctness of the postulate.

The system at one gin was capable of recovering 10-15% of the available heat, while that in the other gin recovered about 40%. With 10-15% heat recovery, the crop was adequately dried under near-ideal conditions. When the crop was very moist, or when ambient relative humidity was high and ambient temperature was low, insufficient drying heat was available at the gin with the low-efficiency heat exchanger. Our data from the gin with the high-efficiency heat exchanger led us to believe that sufficient heat could be supplied to the dryers under almost any conceivable situation. The cost aspects of this source of drying heat are favorable, especially when the alternative fuel is LP gas and the annual production volume at the gin is higher than average.

A. The Ginning Process

The term "seed cotton" is used to describe harvested cotten before it has been ginned. Ginning is the action of separating the fiber from the seed to which it is attached. During the ginning process, seed cotton is dried, cleaned and ginned, in that order. Further cleaning of the lint is done immediately after ginning. The lint is packaged in nominal, 480-lb bales. The bale is the unit on which production, performance, heat needs, and other inputs and outputs are based.

The average yield per acre in the United States is about 1.0 bale with yields of 2.5-3.5 bales being common in irrigated areas and yields as low as 0.5 bale being common in some dry-land areas.

Gins vary in ginning rate from 6 to 40 bales per hour, depending on the equipment used. The annual production volume at gins varies from less than 2,000 bales to more than 30,000 bales, and the operating time per year varies from 200 to 1200 hours. Cotton is ginned as soon as possible after being harvested.

Drying is often done in two stages, between which a cleaning operation is performed. After the second drying stage, further cleaning is done. Cleaning of wet cotton is difficult, so is ginning of wet cotton [1]. On the other hand, ginning at very low moisture content causes fiber damage [1].

B. Gin Waste

Ginning waste is of two types. One is the leaf fragments, sticks and other plant parts removed before ginning, and the other is linty material. Some linty material (known as motes) is rejected during actual ginning, and the remainder is lint removed with foreign matter taken out by the lint cleaner after ginning. At some gins the two types of waste are kept separate and the linty material is sold as "motes," while the non-lint material is the true waste. Other gins discard both components as waste.

C. Objectives

The objective of the study reported here was to collect information needed to design methods of supplying drying heat from the heat released when the waste is burned. Questions posed at the outset were:

(1) Can all the drying heat be supplied from the waste?
(2) Under what conditions is there a deficit or an excess of heat available?

D. Previous Studies

Griffin [2] found that the high heat value of ginning waste averaged 7928 Btu/lb for spindle-picked cotton in the mid-South. The amount of dry waste per bale was 84 lb for first-harvest cotton and 185 lb for second-harvest cotton. The motes were included in these estimates. California is a cotton producing area where motes are often kept separate from other waste and are marketed. One gin company manager [3] reports waste production of 140 lb/bale and mote production of 25 lb/bale (moist basis). Where cotton is harvested with strippers, 750 lb of waste are produced per bale during ginning.

The heat requirement for drying cotton can be derived from fuel consumption data published by Holder and McCaskill [4].

These data translate into heat requirements of about 348,000 Btu/bale when the low heat value and 90% [5] combustion efficiency are assumed. The heat released by burning the waste from one bale would vary from 664,500-1,468,000 Btu. A suitable heat exchanger would thus be one that could extract 24-52% of the flue-gas heat.

McCaskill and Wesley [6] estimated that a 30%-efficient heat exchanger would extract enough flue-gas heat for almost all drying needs and designed their system accordingly. They never operated their incinerator, which was an experimental installation, at a commercial cotton gin. The two incinerators, from which the data presented here were collected, operated at commercial gins and are commercially available, though undergoing design changes shown by experience to be needed.

II. EXPERIMENTS

Our study was started in 1975 at one gin in Arkansas; it was continued in 1976 at that same gin and expanded to include another gin in California.

In 1975, Pitot-static probes and thermocouples were permanently installed in the Arkansas gin. Their output was recorded by multi-point recorders and permitted calculation of the mass air flow rate and heat flow rate to each dryer. To calculate heat flow, ambient conditions were taken as a baseline and the heat flow was the heat added to the air to raise its temperature above ambient.

The results of the 1975 tests have been published [7]. Heat supplied by the incinerator heat exchanger averaged 115,538 Btu/bale and the average seed cotton moisture content was 15.66% (w.b.).

Waste production per bale averaged 136 lb (dry matter) and ranged from 84 lb/bale to 185 lb/bale. The average waste per bale contained combustion heat of 948,823 Btu.* The heat recovery efficiency was thus about 12%. Gas was sometimes used to supplement the recovered incinerator heat. The average gas consumption was 50 ft^3/bale with a heat value of 44,550 Btu/bale.

In 1975, therefore, approximately 72% of the annual heat needs were supplied from the incinerator. This included periods when gas was used to supply all the heat because the incinerator was not working, as well as periods when gas was not used or was unavailable for supplying any heat. Except for start-up from cold each morning, virtually all the heat needed was available from the incinerator.

When the experiment was expanded in 1976, our experimental methods were changed. Instead of using permanently placed instruments, we used a hand-held Pitot-static tube to which a thermocouple was attached, and we made 20-point scans in each air duct in accordance with standard practice [8]. The flow rate was calculated for each scan point and summed over the cross section. The effects of temperature, of static pressure in the duct, and of barometric pressure (but not relative humidity) were considered in our calculations of air-flow volume and mass.

Heat flow rate was the mass air flow rate multiplied by the heat needed per pound of air to produce the temperature difference between ambient air and the air in the duct. This heat-flow rate was thus directly comparable with the amount of heat flowing to the dryer from the gas burners. Four tests, each of several hours' duration, were made on four different dates in Arkansas. Five similar tests were made in California.

Fig. 1 is a schematic drawing of the installation in Arkansas. Air flow and temperature measurements were made upstream of the temperature control system for each dryer. Fig. 2 is a schematic of the California installation. Air flow and

---

* Allowing for 12% non-combustible, soil-derived material.

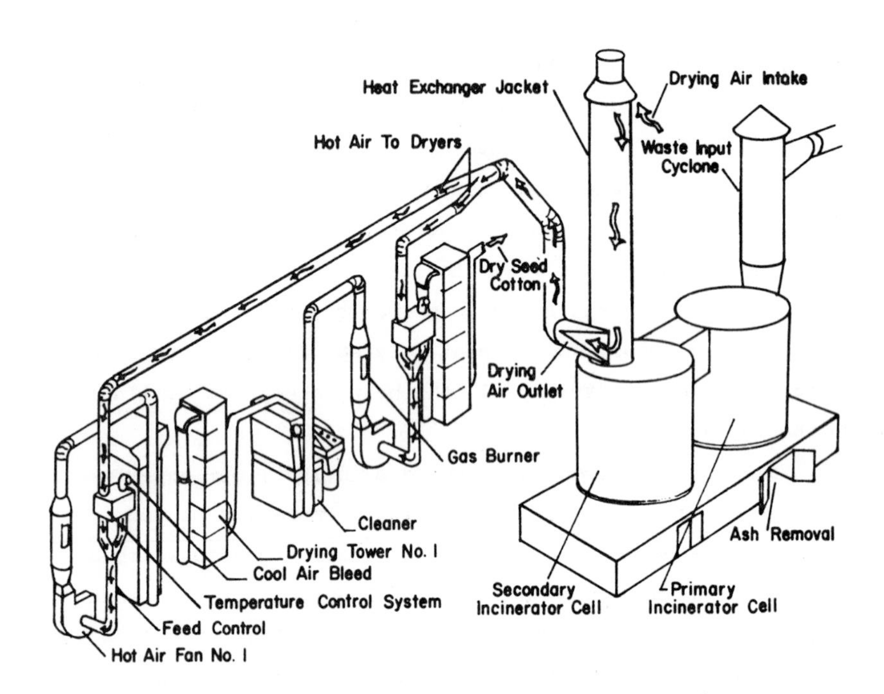

FIGURE 1. Schematic representation of the gin-waste incinerator in Arkansas.

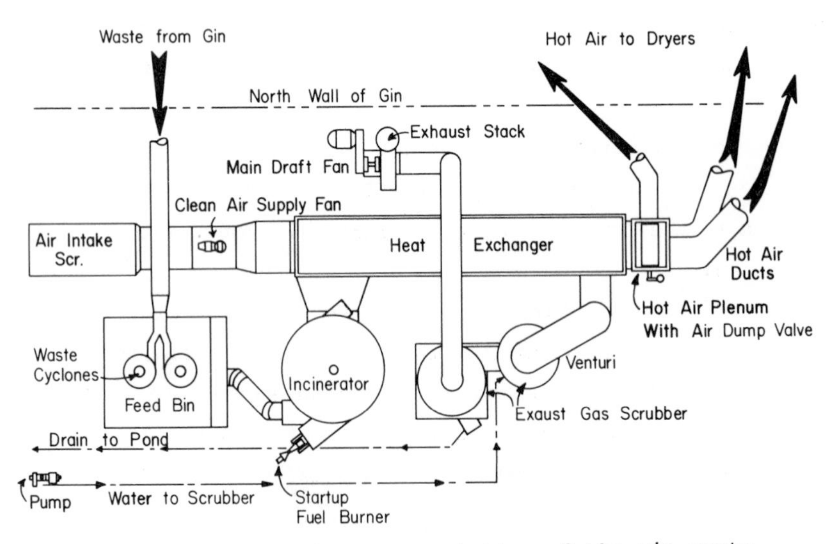

FIGURE 2. Schematic representation of the gin-waste incinerator in California.

temperature measurements were made in each duct leading to the dryers at the inlet to the clean-air side of the heat exchanger. Hot air not used by the dryers was vented. The amount of air vented was calculated from the flow-continuity principle.

III. RESULTS

A. Heat Recovery

Results of the 1976 observations in Arkansas are given in Table I. Gas-derived heat was from natural gas with 990 Btu/ft^3 heat content (low heat value and 90% combustion efficiency). Dryers in a cotton gin are direct fired.

TABLE I. Heat Supplied to Arkansas Dryers

	Drying heat (Btu/bale)		
Date	*From incinerator*	*From gas*	*Total*
10/14/76	*68,074*	*23,636*	*91,710*
10/15/76	-	*105,188*	*105,188*
10/19/76	*80,915*	*93,405*	*174,320*
11/ 5/76	*103,750*	-	*103,750*
Average	*84,246*	*74,076*	*118,742*

The percentage of total heat requirements supplied by the incinerator varied from 100 to 46.4 for days when the incinerator was operating. The incinerator was not operating on October 15, and the supply of natural gas was disconnected on November 5 because of cold weather.

Results of observations in California are shown in Table II. Whenever the incinerator operated, it supplied all the heat needs of the dryers. The heat transmitted from the incinerator to the dryers was derived partly from the heat produced by the start-up burner, which was set to ignite automatically when incinerator temperature dropped below some pre-set level. This heat is not shown separately in Table II as having been derived from gas. In any case, only a small part of the heat transmitted through the heat exchanger to the dryers is derived from gas burning in the incinerator start-up burner.

TABLE II. Heat Supplied to California Dryers

	Drying heat (Btu/bale)		
Date	*From incinerator*	*From gas*	*Total*
11/ 9/76	*378,534*	-	*378,534*
11/11/76	-	*283,872*	*283,872*
11/17/76	*414,403*	-	*414,403*
12/ 6/76	-	*299,216*	*299,216*
12/ 9/76	-	*360,594*	*360,594*
Average	*396,469*	*314,561*	*347,324*

The gas consumed by various processes in the California gin is shown in Table III. The incinerator manufacturer was experimenting with different combustion temperatures in the incinerator. We believe that the gas heat consumption per bale to operate the humidifier and ensure good combustion in the incinerator will be approximately the same as it was on November 9 (61,380 Btu/bale) when the system operates under normal production conditions.

TABLE III. Gas Heat Used by Gin Components (Btu/bale).

	Component			
Date	*Humidifier*	*Incinerator start-up*	*Dryers*	*Total*
11/ 9/76[a]	*46,035*	*15,345*	-	*61,380*
11/11/76	*46,035*	-	*283,871*	*329,907*
11/17/76[a]	*53,703*	*92,070*	-	*145,773*
12/ 6/76	*46,035*	-	*299,216*	*345,252*
12/ 9/76	*46,035*	-	*360,597*	*406,632*

a. *Incinerator operating.*

B. Energy Savings

The data in Table III make possible an estimate of the energy saved by the California system. If the gas consumption on November 9 is taken to be representative of what can be expected, then comparison with consumption on November 11 and December 6 and 9 shows a possible savings of 268,527 to 345,252 Btu/bale can be expected.

The gas-derived heat consumption avoided by using the system in Arkansas is difficult to estimate but is probably about 95,000 Btu/bale. (The average of our tests was 84,246 Btu/bale.) In any case, the total heat used for drying in the Arkansas gin was about half of the 348,300 Btu/bale to be expected from the results of the survey by Holder and McCaskill [4]. On cool, wet days, especially when seed cotton was damp, we recorded heat consumption in excess of 380,000 Btu/bale for short periods at this gin. When crop and weather conditions were ideal, we observed drying with heat consumption of 71,000 Btu/bale.

The drying-heat controls used in cotton gins are less than satisfactory because the heat input rate is not directly related to the moisture to be removed. The reason for this lies in the difficulty of sensing the moisture content of seed cotton, which is a mixture of materials differing in moisture content by as much as 15% within one sample. The drying heat applied to the crop is largely a result of the ginner's judgment. This judgment is based on the condition of the incoming seed cotton, on the condition of the cleaned seed cotton just before ginning, and on the cleanliness of the lint being packaged in the bale.

Resistance-type portable moisture meters are often used to check the lint moisture of seed cotton just before ginning. Some ginners believe the gin does an adequate job with 7 or 8% lint moisture indicated by the meter, others believe the crop should be dried so that only 3% moisture is indicated by the meter.

Seed cotton that is too moist is difficult to clean and sometimes chokes the cleaning equipment. When this happens, production time is lost and severe inconvenience results. On cool, wet days, when cotton may have been rained on in wagons, these choke-ups are common unless the dryer is set to dry the dampest cotton found in a wagon, which may only be a small fraction of the total. This dryer setting might go unchanged for several hours after the damp cotton has been ginned, resulting in a very high heat consumption to dry cotton that is in near ideal condition coming into the process.

The marked difference between the heat consumed in the two gins (about 190,000 Btu/bale) is not attributable to the moisture content of the incoming seed cotton. It is principally a result of the belief at the California gin that higher quality lint that has been thoroughly dried and cleaned will bring higher revenues that more than pay for higher drying expenses. Table IV shows the average moisture contents observed in the two locations. The gin-stand moisture content reflects the moisture content of the seed. Measured with the portable meter, the lint moisture was about 3% at the gin stand.

TABLE IV. Moisture Content of Seed Cotton and Components

	Seed Cotton					
	Wagon	*Gin stand*	*Lint*	*Seed*	*Motes*	*Waste*
Arkansas	*13.46*	*11.88*	*6.19*	*13.56*	-	*19.42*
California	*8.43*	*6.66*	-[a]	*8.20*	*6.47*	*6.83*

a. Lint was rehumidified after ginning.

C. Turnout Analysis

Table V shows the turnout analysis which gives the amount of dry waste available per bale for heat generation. The average is shown although we do not believe we have enough data to give a good estimate of the average waste. Besides, the range over which the waste weight varies is of greater importance than the average for most incinerators--seven of the eight systems we know about are direct-on-line. A direct-on-line system takes the fuel at whatever rate it is available--hence the importance of the range. The alternative is to use a surge bin and to fuel the incinerator at a constant rate at least over periods as short as two to three hours. Even then, the range is important in designing the surge capacity of the system. The California system had a surge bin; the Arkansas system was direct-on-line.

The waste weights shown in Table V are calculated, not directly measured. All other weights given were directly measured. The weight of waste is the weight of seed cotton less the weights of the other components. The slight error caused by dust loss at various points in the gin is believed to be unimportant. However, ginners often calculate the waste weight in a similar manner but based on wet rather than dry waste. Moisture removed by the dryers is thus counted as waste and an overestimate of

Table V. Turnout Analysis (1976)[a]

Date	Seed Cotton	Lint	Seed	Motes	Waste
		Arkansas[b]			
10/14/76	*1323*	*454*	*703*	-	*161*
10/19/76	*1256*	*449*	*701*	-	*106*
11/ 5/76	*1245*	*450*	*695*	-	*100*
Mean	*1275*	*451*	*700*		*122*
		California			
11/ 9/76	*1396*	*460*	*792*	*26*	*117*
11/11/76	*1388*	*455*	*801*	*29*	*103*
11/17/76	*1463*	*455*	*785*	*29*	*195*
12/ 6/76	*1413*	*463*	*782*	*30*	*139*
12 9/76	*1501*	*460*	*779*	*26*	*236*
Mean	*1430*	*459*	*788*	*28*	*158*

a. Units are lb dry matter per 480-lb bale of moist lint.

b. The range of waste weights in 1975 was 84-185 lb/bale.

waste weight always results. Using the waste moisture content as an adjustment factor is incorrect.

D. Percentage Heat Recovery

From Table I, the average heat recovered per bale at the Arkansas gin was 84,246 Btu. Assuming a heat value of 6986 Btu/lb (after allowing for soil contamination), the dryers receive about 10% of the available heat. This is consistent with 1975 data.

From Table II, the average heat delivered from the incinerator to the dryers in the California gin was 396,468 Btu/bale and the average waste available was 158 lb. Bomb-calorimeter determinations yielded an average heat content of 6225 Btu/lb. The dryers therefore received an average of 40% of the available heat.

E. Non-Combustibles

High soil-contamination levels were found in the California waste, up to about 20% of the sample dry weight. This accounts for the lower heat value than was reported by Griffin [2]. A similar situation existed in Arkansas, but we have no data to show the contamination level. We assumed that about 12% would be soil-derived, non-combustible material.

The incinerator in Arkansas was designed to accumulate ash in the primary cell, which was cleaned out daily. But clinkers and "glass" formed in the primary cell and often made clean-out difficult. Ash encrustations often blocked the free entry of combustion air.

The incinerator in California was of a design that required virtually all non-combustible matter to be carried along with the combustion gases, through the heat exchanger, and into a wet-venturi scrubber which removed them from the stack gas. Because the levels of soil contamination were higher than had been expected, some soil-derived non-combustible material dropped out of the gas stream and accumulated in the combustion chamber and in other places where gas velocity slowed due to a widening cross section. The accumulation eventually restricted air flow through the combustion space and this caused the combustion-space temperature to increase to a point at which the incinerator had to be shut down.

Ideally, the incinerator should be designed to cope with high levels of soil contamination, and the manufacturer of the California incinerator now believes that the necessary changes have been made to accomplish this ideal. Another approach would be to remove as much of the undesirable material as possible from the waste before incineration. Methods of accomplishing this are now being studied. Virtually all the heat-recovering incinerators at cotton gins have been adversely affected by the high non-combustibles content of the waste material.

F. Cost Aspects

When the California incinerator was operating, LP gas consumption of about four gallons per bale was avoided--this was the gas consumption rate of the dryers when the incinerator was not operating. However, the start-up burner in the incinerator can be expected to consume about 0.25 gallon per bale when the incinerator is operating, resulting in a net saving of about 3.75 gallons per bale. The cost of LP gas varies from place to place and depends on quantities purchased in a single consignment, but $.36 per gallon is representative. The expense saved would then be about $1.35/bale, or about $40,000 per year when annual volume is 30,000 bales. These savings would make available the revenue needed to accomplish capital recovery on $160,000 with 10% compound interest in about five years. When applicable tax reductions resulting from investment credit and depreciation allowance, and other provisions (such as avoided disposal cost) are considered, additional net revenue would be available. However, labor, repairs, increased insurance (if any), and increased property tax would be expenses to be met from some of the available revenue.

Natural gas was the fuel used in the Arkansas gin, and it cost $1.75/MCF in fall 1976. Gas-derived heat use avoided by using the incinerator in Arkansas is estimated at about 95,000 Btu/bale, or 107 ft^3 of gas if combustion is assumed 90% efficient. This is a saving of 19 cents per bale. For cotton gins in the mid-South as a whole, natural gas costs for drying are about 50 cents per bale. As discussed earlier, gas consumption is often a function of the ginner's judgment. The survey referred to [4] resulted in a heat consumption estimate of 348,300 Btu/bale. The Arkansas gin used 165,000 Btu/bale in 1974, before the incinerator was installed. If the Arkansas gin could fully eliminate gas consumption, it would therefore hope to save 33 cents per bale at 1976 prices for natural gas.

We estimate that the Arkansas incinerator and associated equipment would cost $100,000. Revenue of about $25,000 per year is needed for capital recovery in five years with 10% interest compounded annually. This means that an annual volume of about 75,000 bales would be needed--something which is clearly impossible at this gin where maximum volume of 20,000 bales per year is possible but unlikely. Tax savings resulting from various deductions related to the equipment, and savings due to not having to continue paying for waste disposal would free additional revenue to contribute to capital recovery. Disposal costs alone are often 75 cents to $1.00 per bale. When such aspects of the situation as unavailability of natural gas or LP gas are considered, using the incinerator to generate drying heat may be the most economical option. This situation had already arisen before the decision to install the Arkansas incinerator was made. Nevertheless, the conclusion is unescapable that gins at which LP gas is the fuel used are the ones most likely to benefit from installation of a heat recovering incinerator. Circumstances such as the frequent unavailability of natural gas and/or high alternative disposal costs could justify the system at gins now dependent on natural gas.

IV. CONCLUSIONS

The equipment used at the California gin demonstrates that enough heat for drying cotton under almost any circumstances can be recovered from burning gin waste. Approximately 40% of the available heat was supplied to the dryers in the California gin.

At the Arkansas gin, where only about 10% of the heat was recovered for drying, gas was often used as a supplementary heat source. The amount of gas used in this way could be kept at very low levels by careful operation of the gin, but a 10% heat recovery level did not give the flexibility needed to cope with difficult drying situations.

Investment of capital in heat recovering equipment is most likely to be justified when the combined cost of purchasing fuel and disposing of waste exceeds $1.50 per bale.

ACKNOWLEDGMENT

The cooperation of Robert G. Curley, George E. Miller and O. D. McCutcheon of the University of California Cooperative Extension is acknowledged.

REFERENCES

1. Handbook for Cotton Ginners. Agriculture Handbook No. 260, USDA, ARS, 1967.
2. Anselem C. Griffin, Jr., "Fuel Value and Ash Content of Ginning Wastes," Transactions of the American Society of Agricultural Engineers 19(1), 156-158, 1976.
3. Macon Steele, Producers' Cotton Oil Co., Fresno, California. Private communication, 1976.
4. Shelby H. Holder, and Oliver L. McCaskill, "Costs of Electric Power and Fuel for Dryers in Cotton Gins, Arkansas and Missouri." ERS 138, USDA, ERS and ARS, 1963.
5. A. W. McConnell, McConnell Sales, Birmingham, Alabama. Private communcation, 1977.
6. O. L. McCaskill and R. A. Wesley, "Energy From Cotton Gin Waste." The Cotton Ginners Journal and Yearbook, 44(1), 5-14, 1976.
7. W. F. Lalor, J. K. Jones, and G. A. Slater, "Performance Test of Heat-Recovering Gin-Waste Incinerator." Cotton Incorporated Agro-Industrial Report 3(2), Raleigh, NC, 1976.
8. Industrial Ventilation, 14th Ed. American Conference of Governmental and Industrial Hygienists, Lansing, MI Section 9 1976.

THE DESIGN OF A LARGE-SCALE MANURE/METHANE FACILITY

Frederick T. Varani and John Burford

Bio-Gas of Colorado
Arvada, Colorado

Richard P. Arber

CH2M HILL, Inc.
Denver, Colorado

I. INTRODUCTION

Since the advent of the energy crisis and subsequent accelerating natural gas prices, interest in obtaining substitute natural gas from alternate sources has been widespread. One area of interest, for example, has been the gasification of coal and oil shale. Many dollars have been invested in this research area; and several projects have been contemplated and designed. However, the reported cost estimate of $1 billion for a gasification plant has kept all of these projects on the drawing board to date.

Interest also has been widespread in creating substitute natural gas from the so-called "bio-gasification" of organic materials. Bio-gas plants based on anaerobic fermentation could be constructed on a more modest scale than coal or oil shale pyrolysis facilities and still (as it appears to many people in this field) produce commercial quantities of substitute natural gas at costs which are acceptable in view of the natural gas shortages.

Institutions such as the University of Illinois, Cornell, and others are supporting research on this source of gas. The Institute of Gas Technology has been operating anaerobic digestion apparatus for several years. Also, many agricultural colleges currently are designing or operating digestion systems for animal wastes, while many private companies are heavily involved in the promotion and implementation of the bio-gasification idea. Furthermore, ERDA has issued a Request for Proposals for a special manure/methane facility to be built and operated as an experimental facility. The proposed ERDA project, based on work by Dynatech Corporation [1], is to determine the commercial value of providing substitute natural gas by the anaerobic fermentation of manure.

All these activities indicate that fermentation of animal manures as a substitute natural gas source is a technical concept moving rapidly from laboratory research into demonstration and even commercial operation. For example, Thermonetics (formerly CRAPCO) has announced that its own privately financed manure/methane facility will go on-line in the fall of 1977. However, little has been published about the design of this plant. This paper will deal with the evolving design of a large-scale anaerobic digestion facility to produce methane gas being proposed for the City of Lamar, Colorado.

II. LAMAR FACILITY BACKGROUND

Since 1973, Bio-Gas of Colorado, Inc., has been involved in the application of anaerobic digestion for the production of fuel gas from organic waste materials. They are now operating a 6000-gallon mobile digestion unit and have been working on designing large-scale facilities for the production of pipeline quality gas. The work has been supported by grants from the 4-Corners Regional Commission.

Recently, a project team composed of Bio-Gas of Colorado, Las Alamos Scientific Laboratories, the City of Lamar Utilities Board and CH2M HILL, Inc., has been formed to complete a preliminary design of a full-scale facility for the City of Lamar. This proposed facility would utilize the manure from 50,000 feedlot cows in the Lamar area as a feed source for the anaerobic digestion plant. The resulting fuel gas would be consumed in the Lamar Utilities Board's electrical generation plant.

The Lamar site was one of many potential sites for a bioconversion facility identified by the project team [2]. Contractual arrangements were made between the local area feedlots and the Lamar Utilities Board which have allowed this project to move from the idea stage into one of implementation.

III. PILOT PLANT

To collect data and test various processes and unit operations being considered for the full-scale facility design, a pilot facility was constructed and is currently in operation. This pilot plant, shown on Fig. 1, consists of a heated 200-gallon (757 liter) mixed, main reaction vessel and a 200-gallon unmixed, unheated reaction, second stage vessel. A slurry of manure is fed into the pilot plant on an hourly basis. Roughly, 5 pounds of organic solids per day are processed, and 20 standard

FIGURE 1. Bio-gas pilot plant.

cubic feet of methane per day is generated. The pilot plant is also used to generate residue materials to be processed in centrifuges and other unit operations proposed for the full-scale facility design. Table I shows some basic average daily operaing characteristics for this plant.

TABLE I. Typical Pilot Plant Operating Characteristics

Feed rate	*4.87 (2.21 Kg) lbs vs/day,* *9.46 gal (35.81 liters) 24 hrs at* *9% solids*
Manure analysis	*55% solids* *66% volatile solids on dry basis*
Mixing rate	*1 min. out of every 30 min.*
Retention time	*20 days*
Digester temperature	95°F (35.0°C)
Gas production rate	*37 1 ft^3 (1046 liters) bio-gas/24 hrs* *at 35% by volume Co_2* *at Denver conditions (630 MM hg and* *20°C*
Yield	*(4.05 scf CH_4/lb v.s. added)*

IV. DESIGN CRITERIA FOR A FULL-SCALE FACILITY

Late in 1976, CH2M HILL, a national wastewater treatment consulting engineering firm, was contracted to do the preliminary full-scale design for the Lamar facility. After evaluating unit operations, CH2M HILL, with Bio-Gas, established the design criteria for the full-scale facility, as shown on Table II. The selected flow schematic which was evolved from the research done to date is shown on Fig. 2. Basically, the full-scale plant will have the following four major components:

(1) Mixing and grit removal facilities to produce a manure slurry and remove sand particles
(2) A digestion complex consisting of anaerobic digesters to process the slurry and produce methane gas, heat generation from the power plant, and CO_2 removal facilities
(3) A degasification tower and centrifuge to process the sludge from the digesters for disposal
(4) Algae ponds to treat the effluent from the centrifuges for recycling through the plant

TABLE II. Full Scale Design Criteria

Number of cattle supply manure (design point)	*50,000*
Annual manure load tons/year	*125,000*
Total gas produced ft^3/CH_4 at STP conditions 24 hrs	*1,226,450*
Design plant loading rate lbs volatile solids/ft^3 digester	*.20*
Volatile solids destruction	*55%*
Digester temperature	*95°F*

In addition to recycling the process water, the plant will have several other recycling systems as diagrammed on Fig. 2 and discussed in the detailed descriptions of the major plant components. Before these details are presented, however, the following paragraphs discuss some of the factors that have been key considerations in the development of the design criteria.

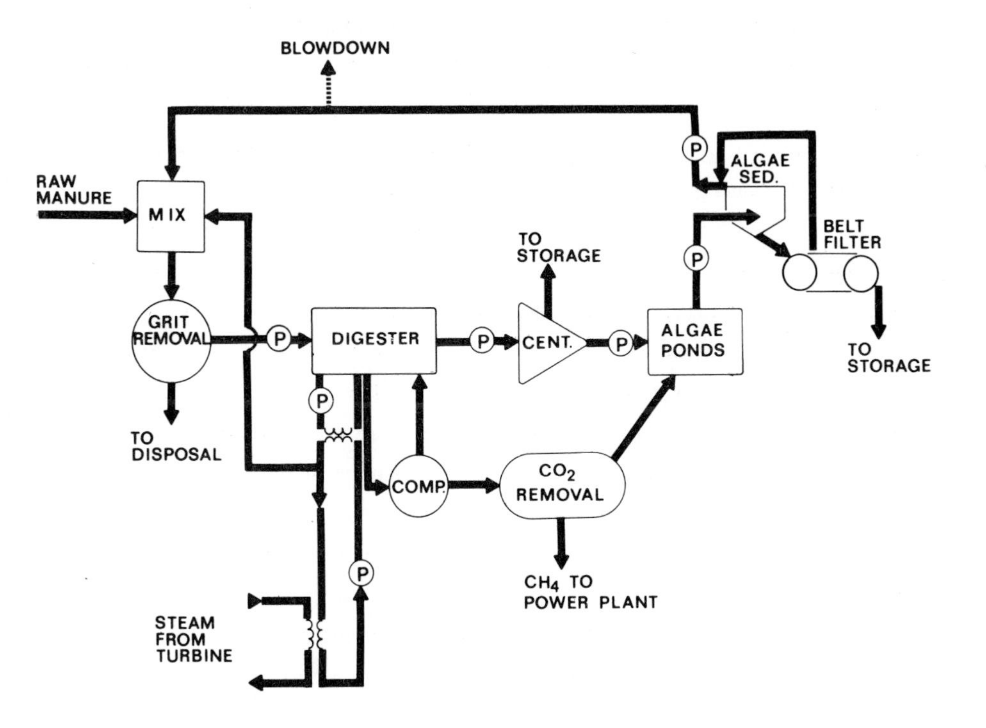

FIGURE 2. Selected flow schematic.

One significant design consideration is the large amount of solids that the bio-conversion facility will produce for separation and disposal. The solids will vary from 75 to 145 tons per day on a dry weight basis. By way of comparison, the proposed solids disposal system for the Metropolitan Denver Sewage Treatment Plant is being designed to handle 110 tons of dry solids per day. Thus, the solids handling capability of the bio-conversion facility will be comparable to that required to handle the solids for a city of approximately 1 million people.

Another major design factor for this bio-conversion facility is the objective of recycling as much water as possible, thereby reducing the consumptive use of water. This is particularly important since the facility will be located in the semi-arid region of eastern Colorado. A complete recycle system with no consumptive use of water appears to be impractical because several toxic elements may build up to a level which may inhibit the gas production in the anaerobic digester. Although, as discussed later, it is possible that the algae ponds will assimilate these toxic elements, a blowdown system to reduce the effluents toxic content is being designed as part of the facility until the capabilities of the algae ponds are monitored.

The blowdown system must remove enough water to prevent the buildup of toxic materials to a level that would reduce the gas production. A preliminary analysis of the amount of blowdown required indicates a consumptive use of approximately 100 to 200 acre-feet of water per year. Presently, the design calls for the blowdown water to be evaporated in a pond located onsite, thereby eliminating discharges to any of the surface streams. The recycled water must be treated to remove ammonia, sodium, calcium and magnesium, which would otherwise build to toxic levels in the recycle stream.

Perhaps the most unique feature of the Lamar manure-gas plant is the mini-algae farm/water recycle unit. The Los Alamos Scientific Laboratory has received a parallel grant from the 4-Corners Regional Commission to design an algae growth system using the centrifuge effluent as the growth media. The interesting concept employed in this component is the possibility of the algae assimilating the potentially toxic materials. The algae will be harvested as a protein source, and the water minus the toxic elements will be recycled through the system.

The temperature in the digester facilities is one of the most important design criterion. The anaerobic fermentation process must be temperature controlled at the selected operating temperature of 95°F to ensure an optimal volatile solids destruction and gas production. In order to produce gas successfully, the two processes of generating power and waste heat and producing bio-conversion gas must be carefully integrated and considered as a unit.

Variations in heat loads from the power plant will have a significant impact on the production of gas in the bio-conversion facility. Similarly, variations in the quantity of gas produced and its Btu value could also have a significant impact on power generation when it is recycled back to the power plant. The gas purification system for removing carbon dioxide from the digestor gas must increase the Btu value from less than 500 Btu/ft^3 to a value which, when combined with the natural gas supply to the power plant, results in at least 850 Btu/ft^3.

V. MIX AND GRIT REMOVAL FACILITIES

Dump trucks will haul raw manure from area feedlots to the bio-conversion facility site. The trucks will be weighed and the volatile solids content of the manure determined. The manure will be stored in piles surrounding the mix tanks. A maximum of ten days storage will be provided. Tarpaulins will cover the stored manure to help reduce odors.

Front end loaders will be used to dump the stored raw manure into a hopper unloading onto two belt conveyers. The belt conveyors will transport the raw manure to two concrete 2000-gallon mix tanks. A scale beneath the conveyors will record the raw manure feed. Each of the tanks will have a 20-horsepower mechanical mixer to blend the raw manure with water into a slurry of approximately 10% solids. The high horsepower to volume ratio is required to break the clumps of manure as well as to keep the grit suspended. The mix tanks will provide approximately 10 minutes of detention time.

The mixture of manure and water will then flow through baffles to two circular grit chambers where sand particles greater than 65 mesh will settle and be removed from the solution to prevent damage to downstream equipment. The grit will be scraped to a sump, conveyed from the basin by an auger, washed, and dumped into a truck for disposal off site. The degritted slurry will then be pumped to the digesters by four progressive cavity pumps.

VI. DIGESTION COMPLEX

The digestion complex for producing the methane gas will consist of four anaerobic digesters and a control building, a heat exchange system, and gas handling and conditioning facilities. Each of these components is discussed below.

A. Anaerobic Digesters

Four digesters have been proposed for this full-scale plant. They will be earthen basins, 135 ft by 135 ft with a 30 ft sidewall depth. The basins will be constructed with reinforced earth retaining walls, which are one of the innovative design features being planned at this time.

The retaining walls will be covered with a plastic membrane to make them watertight. The floors of the digesters will be asphalt. The covers will be constructed of precast concrete double tees with a 2.5 in. concrete top to make them gastight. The digester feed, drawoff and gas mix piping will be supported by the roof. The gas will be maintained at a slight positive pressure beneath the covers.

The raw sludge feed from the grit chamber will be measured by magnetic flow meters. Each digester will be provided with gas compressors, diffusers and controls for gas mixing. The digesters will be completely mixed. A positive displacement drawoff pump for each digester will be provided.

A digester control building will contain controls and equipment for feeding the raw sludge, compressing the gas for mixing, and drawing off the digested sludge. Several operational parameters will be measured for each digester, including temperature, pH, alkalinity, volatile acids of the sludge, gas production and percent CO_2. The digester control building will also contain offices, a laboratory, central motor control center, showers and restrooms.

B. Heat Exchange System

The Lamar power plant will provide the heat source for the digestion process. Discharged water from the power plant's condensor at 85°F will be heated with steam from extraction points on the turbine in a heat exchanger. The condensate will be heated to approximately 200°F and pumped to the bio-gas facility through insulated pipes.

The hot water from the power plant will pass through spiral heat exchangers where the heat will be transferred to the sludge to raise the sludge temperature to 95°F. After leaving the heat exchangers, the water is expected to be approximately 145°F. Some of this water will be used for heating the algae ponds and providing makeup water for the mix tanks. The balance of the water will be returned to the power plant, where it will be reheated and recirculated.

An investigation was performed to determine whether the bio-gas should be processed to a quality acceptable for use in the existing power plant facilities or whether the existing boilers in the power plant should be converted to accept untreated bio-gas.

It was found that conversion of the existing boilers would require additional manpower in the power plant. Also, the capacity of the existing boilers did not provide a good match with the amounts of bio-gas to be produced. Therefore, it was decided that carbon dioxide should be removed from the bio-gas prior to use.

C. Gas Handling and Conditioning

The bio-gas will be withdrawn from the digesters and compressed. As mentioned above, the CO_2 will have to be removed from the gas for use in the power plant. The removal process has not yet been selected. However, it is planned that the CO_2 will be used in the algae ponds for pH control and mixing. Several carbon dioxide removal processes are being investigated. It appears that a water absorption process has the most merit.

VII. DEGASIFICATION AND DEWATERING SYSTEMS

The homogeneous sludge from the digesters will be pumped to a degasifying process to make it easier to dewater. The sludge will be introduced into the top of a 30 in. diameter by 30 ft high enclosed tower, where it will impinge on a plate to create a spray. This spray will aid in releasing gas from the sludge. Two vacuum pumps will withdraw the released gas from the top of the tower for discharge to the atmosphere.

Two solid bowl dewatering centrifuges, each with a 175 gpm capacity, will be used to dewater the digested sludge. A polymer will be added to the sludge prior to centrifuging to condition it for dewatering. The centrifuges will concentrate the sludge to approximately 25% solids and are expected to have between 95% and 99% solids capture, as has been demonstrated by the pilot testing.

The dewatered solids (cake) from the centrifuge will be conveyed from the centrifuge building by a belt conveyor. The solids will be deposited into a stacker conveyor, which will transport the solids to an onsite storage area. Six months of storage will be provided on the site.

It is expected that the stored solids will dry to a moisture content of approximately 50% during storage. The dewatered solids will be used as a fertilizer by the area farmers or as a supplement to cattle feed. The suspended solids-free centrate will be pumped from the centrifuge building to algae ponds for processing.

VIII. ALGAE PONDS

The algae ponds will consist of 6 acres of concrete basins, 18 in. deep. Three species of algae will be grown in the ponds: Anabena, Spirulina, and Chlorella. Detention time of these ponds will be approximately 5 days. The temperature of the water will be approximately 85°F and the pH, approximately 7.8.

The ponds will be covered by conventional greenhouse structures utilizing a double layer of 6 mil. transparent polyvinyl chloride. The basins will be mixed to permit the algae to come to the surface to be exposed to the light. The pH will be adjusted by injecting CO_2 into the liquid. As noted above, the CO_2 source for mixing and pH adjustment will be from the removal facility for the digested gas.

The ponds will be harvested after the 5-day detention, using a settling tank with detention time of approximately 8 hrs. The settled algae will be concentrated to approximate 5% to 6% solids. A belt filter will be used to dewater the settled algae. The liquid from the dewatered algae will be recycled back to the mix tanks. The ponds will be closely monitored to assess their capability to remove toxic materials from the effluent, as discussed earlier. Meanwhile, blowdown of the flow returning to the mix tanks is planned.

IX. COST ESTIMATES

The cost estimates for the proposed facility have not yet been fully developed. However, a preliminary estimate indicates that the capital costs could range from between $7.5 and $10 million, and operation and maintenance costs could range from $400 to $700 thousand per year (1977 costs).

X. FUTURE WORK

A great deal of work still remains to be done to complete this full-scale facility design. It is believed that the unit operations selected to date have been proven by the pilot plant or experience, and will allow the facility to operate as envisioned. However, the selected operations may not be the optimum systems either from a capital cost or operation cost standpoint.

The immediate plans are to operate the 200-gallon pilot plant at "off design" conditions to observe the effect on gas production and residue. Bio-Gas will continue to operate the mobile digestion unit at Lamar area feedlots to make enough residue for feeding trials of the protein values and to further refine the design of the various full-scale subsystems such as the centrifuges and algae ponds.

Finally, the design of the large-scale facility will go into the detail design, plans, and specifications phase. It is hoped that the project will receive funding and a go-ahead into construction by early 1978.

REFERENCES

1. Edward Ashare, Donald L. Wise, and Ralph L. Wentworth, Fuel Gas Production From Animal Residue, Dynatech Corp. 1551, ERDA contract EY-76-C-02-2991.*000.
2. J. Burford, Jr., and F. T. Varani. Energy Potential Through Bio-conversion of Agricultural Wastes. Final report to the 4-Corners Regional Commission. Technical Assistant, Grant FCRC 651-366-075.

ENERGY RECOVERY FROM MUNICIPAL WASTES

James R. Greco

National Solid Wastes Management Association
Washington, D.C.

I. INTRODUCTION

Numerous technologies are being developed for extracting energy and resources from municipal solid waste (MSW). Some of these systems are being implemented throughout the United States and Canada. As the cost of fossil fuels rise, it is expected that MSW will appear more economically attractive as an energy resource. As a result, various approaches are now in the research, development, demonstration, and commercialization phases for the preparation of solid, liquid, and gaseous fuels; the production of steam and the generation of electricity.

When determining the national potential for energy production from MSW, analysis often begins with national waste generation estimates and the projected waste generation for particular time periods (e.g., 200 million tons of MSW generated in the U.S. in 1980). Table I presents several such estimates, and updates them to 1975 through the application of various growth rates.

TABLE I. National Waste Tonnage Estimates: References and Projected 1975 Levels

Reference[a]	Base year	Annual tonnage (millions/yr)	Annual tonnage (millions/yr) projected for 1975 for these growth rates (percent/year)					
			0	2	4	6	8	10
1968 National Survey (low)	1967	190	190	223	260	303	352	408
(high)	1967	291	291	341	399	464	538	625
1971 Private Sector Survey	1970	265[b]	265	292	322	355	390	427
1971 EPA RRD estimate	1971	125	125	135	146	158	170	184
1971 Bureau of Mines estimate	1971	129	129	140	151	163	175	189
1973 EPA RRD estimate	1973	135	135	140	146	152	158	164

a. The references above address different waste sources, for example:

(1) the 1968 National Survey represents estimated wastes reportedly for residential (88 million tons), commercial and institutional (57 million tons), and some "low" estimates for industrial (34 million tons), demolition and debris (5 million tons), and other municipal (6 million tons) wastes. "High" estimates for these latter three waste groupings would raise the 190 million ton estimate to 291 million tons per year. The 1968 survey of municipal officials was essentially a mail survey of estimated tonnages rather than measured data [1].

(2) the 1971 Private Sector Survey includes national estimates for residential (119 million tons), commercial (77 Million tons), and industrial (69 million tons) wastes. Residential tonnage represents wastes collected from single family dwellings and apartments of 4 units and less, whereas commercial wastes are from apartments with greater than 4 dwelling units, stores, offices, institutions, etc. Industrial tonnage describes solid wastes from manufacturing and industrial processing

plants. The private sector estimates were developed from interviewing roughly 10% of the total number of private contractors and statistically extrapolating the data to be representative nationwide [2].

(3) the 1971 and 1973 EPA Resource Recovery Division estimates include household, commercial, and institutional wastes derived from material flow studies and product/source composition analysis of municipal solid wastes. Industrial processing wastes were excluded [3,4].

(4) the 1971 Bureau of Mines estimate is described as dry urban refuse and excludes industrial wastes [5].

b. Determined by multiplying the daily tonnage estimate of 878,500 tons per day by 5.78 workdays per week and 52 weeks per year.

The most frequently employed estimate, for 1975, is 135 million tons as projected from the 1971 Environmental Protection Agency study.

An additional potential energy source exists with those wastes previously disposed of in landfills. These landfills are potential reservoirs of methane. This gas can be marketed either to industrial gas customers or upgraded to pipeline quality natural gas. Consequently, energy can be recovered both from presently generated MSW and from previously disposed of wastes [6].

The various technologies developed have been categorized into four generic types: (1) steam generation; (2) pyrolysis; (3) mechanical fuel preparation; and (4) landfill gas recovery. The first three approaches depend upon MSW being delivered to a central processing facility. The last technology is implemented at an existing disposal site.

Processes handling as-generated waste vary in the degree of waste processing required to yield a marketable product. Steam generation generally involves the burning of unprocessed MSW in waterwall incinerators. Pyrolytic and mechanical fuel processes require size reduction of the feed material by shredding or wet pulping. Following this preliminary processing step, the wastes become more homogeneous. Pyrolysis processes then to destructively distill this material while mechanical systems necessitate air or water classification--material separation steps for removing noncombustibles from the fuel product. Numerous fuel preparation processes are being developed and applied: the shred-air classify approach, the more common shred-air classify-secondary shred process, and the refined shred-air classify-secondary shred-chemically treat method.

Because methane gas is produced continuously in a landfill as natural composition of organic waste occurs, the technology developed to date has focused upon collecting and purifying this gas. Extraction through wells is the method employed for gas recovery. The yield of these wells is determined by several factors including the rate of gas production, refuse composition and cover material integrity [7].

The marketable products of these four technical approaches are steam or electricity from incinerators; gas, oil, or steam from pyrolytic processes; solid fuels from mechanical fuel preparation systems; and gas from methane recovery at landfills. Theoretically ethanol, methanol, ammonia and other chemicals can be derived from pyrolytic and landfill gases also.

II. STATUS OF TECHNOLOGY DEVELOPMENT AND APPLICATION

Table II presents a compilation of energy recovery projects in the U.S. and Canada. Projects with at least 100 tons per day (tpd) capacity were included [8]. Other projects of lesser size have undergone research and development. In Tonawanda, New York, the 75 tpd Torrax gas pyrolysis plant was built. Other test facilities exist in Menlo Park, California, and in other cities. These operations have focused primarily on pyrolysis and incineration techniques. Additional projects of less than 100 tpd design capacity include modular combustion units installed in Blytheville, Arkansas; Groveton, New Hampshire; and Siloam Springs, Arkansas.

The design capacities given in Table II are in tpd units. The reader should note that comparisons should be made cautiously. Some plants may operate only for 8 hours or less per day whereas others may employ 16-24 hr/day operations. Furthermore, a plant may have greater design capacity through a parallel system design (redundant process lines). As a result, tpd comparisons may be misleading.

The status of each project listed in Table II conveys whether the facility is in the construction or operation phase.

TABLE II. Centralized Resource Recovery Facilities in the United States

Technology (location)	Design capacity (tpd)	Status	Start-up (year)	Operator	Project characterization	Energy type	Materials
Heat recovery							
Akron, Ohio	1000	Construction	1978	Teledyne National	Commercial	Steam	Fe
Braintree, Mass.	240	Operational	1971	City of Braintree	Commercial	Steam	
Chicago, Ill.	1600	Operational	1972	City of Chicago	Incinerator	Steam	Fe
Harrisburg, Pa.	720	Operational	1972	City of Harrisburg	Incinerator	Steam	Fe
Nashville, Tenn.	720	Operational	1974	Nashville Thermal Transfer Corp.	Commercial	Steam	
Norfolk, Va.	360	Operational	1967	U.S. Navy	Commercial	Steam	Fe
Onondago Co., N.Y.	1000	Construction	1979	Co. of Onondaga	Commercial	Steam	Fe
Saugus, Mass.	1200	Operational	1975	Refuse Energy Sys.	Commercial	Steam	Fe
Pyrolysis							
Baltimore, Md.	1000	Modification	1975	City of Baltimore	Demo unit	Steam	Fe,Glass
So. Charleston, W.Va.	200	Operational	1974	Union Carbide	Pilot Plant	Gas	
San Diego Co., Cal.	200	Construction	1976	San Diego Co.	Demo unit	Oil	Fe,Al, Glass
Mechanical fuel preparation							
Albany, N.Y.	600	Design	1978-9	City of Albany	Commercial	RDF	Fe
Ames, Iowa	200	Operational	1975	City of Ames	Commercial	RDF	Fe,Al, Glass
Baltimore Co., Md.	1200	Operational	1976	Md. Environ. Serv. Teledyne National	Research & Development	RDF	Fe

TABLE II. (Continued)

Technology (location)	Design capacity (tpd)	Status	Start-up (year)	Operator	Project characterization	Energy type	Materials
Bridgeport, Conn.	1800	Construction	1978	CEA/OXY Resource Recovery Asso.	Commercial	RDF	Fe, Al, Glass
Chicago, Ill.	1000	Operational	1977	City of Chicago	Commercial	RDF	Fe
East Bridgewater, Mass.	1200	Operational	1976	Combustion Equipment Associates	Research & Development	RDF	Fe
Hempstead, N.Y.	2000	Construction	1978	Hempstead Resource Recovery Corp.	Commercial	Steam	Fe, Al, Glass
Houston, Texas	400	Operational	1971	Browning-Ferris Industries, Inc.	Research & Development	RDF	Fe
Milwaukee, Wisc.	1000	Operational	1977	American Can Co.	Commercial	RDF	Fe, Al, Glass
Monroe, N.Y.	2000	Construction	1977-8	Raytheon Service	Commercial	RDF	Fe, Glass
St. Louis, Mo.							
U.S. EPA Demo unit	325	Operational	1972	City of St. Louis	Demo unit	RDF	Fe
Union Colliery/ Union Electric	8000	Re-evaluation	1979	Union Colliery/ Union Electric	Commercial	RDF	Fe
Other							
Pompano Beach, Fla.	100	Construction	1977	Waste Management, Inc.	Experimental	Gas	Fe, Al

Source: National Solid Waste Management Association.

Operational here means that the plant is built and can be operated, but it does not necessarily imply daily operation. The start-up year given identifies the year that refuse was or will be initially processed. Project characterization indicates whether the project is in one or another stage of research and development, or whether it is a commercial plant. Chicago, Illinois, and Harrisburg, Pennsylvania are listed as incinerators because the steam being produced at these facilities is not being sold to a customer. Only volume reduction is being accomplished, although customers are being sought and may have been recently secured.

The implementation of these technologies is increasing. Table III presents the number of energy recovery projects by the year in which they started up (or will start operations). The first project was the U.S. Naval Station in Norfolk, Va., which began producing steam for ships in port in 1967. In 1972, the St. Louis demonstration project, partially funded by the EPA, began operations. From 1974 through the present, a number of facilities began operations including both mechanical fuel systems and pyrolysis technologies. Three pyrolysis plants came on line in that period: Baltimore, Maryland (low temperature gas pyrolysis); South Charleston, West Virginia (high temperature gas pyrolysis); and San Diego County (flash pyrolysis yielding pyrolytic oil). Many observers believe that the leveling off in new operational capability forecast for 1977-1979 reflects a "wait and see" posture. Generally it takes 3-5 years to plan and implement a technology-oriented program for recovering energy from waste.

TABLE III. Tabulation of the Number of Resource Recovery Plants Which Became Operational

Technology	Plants started operation													Total
	1967	1968	1969	1970	1971	1972	1973	1974	1975	1976	1977	1978	1979	
Heat recovery	1	0	0	0	1	2	0	1	1	0	0	1	0	7
Pyrolysis	0	0	0	0	0	0	0	1	1	0	0	1	0	3
Mechanical fuel preparation	0	0	0	0	1	1	0	0	1	2	2	4	0	11
Other	0	0	0	0	0	0	0	0	0	0	1	0	0	1
Total	1	0	0	0	2	3	0	2	3	2	3	6	0	22

III. PROCESS DESCRIPTIONS

As noted previously, the energy from waste technologies being developed can be grouped into four major categories. These four generic approaches are explored, in more detail, in this section.

A. Steam Generation

The generation of steam from MSW can be accomplished by burning unprocessed or processed materials. Generally the extent of processing employed depends upon the incorporation of shredding, air classification and materials separation. The Hamilton, Ontario, facility and the Akron, Ohio, project under construction exemplify steam generation plants fed by processed MSW.

Until recently, U.S. design practices for refuse-fed steam generation plants employed refractory furnaces with internal or external boiler tubing. Merrick, New York; Miami, Florida; and Oceanside, New York are examples of this design. Current U.S. trends appear to favor waterwall furnaces developed and deployed in Europe. Waterwall furnace designs have replaced refractory lined furnace designs because the waterwall units are less expensive to purchase, less expensive to maintain, and more efficient in the combustion of refuse and generation of steam. The first commercial application of waterwall technology in America for the burning of MSW is the Norfolk facility previously mentioned. The largest built to date is the facility located in Saugus, Massachusetts. Both systems are shown in Table II. The Saugus plant is described below.

1. Unprocessed Waste Utilization

In the generation of steam from unprocessed waste, MSW is deposited on a tipping floor or in a large storage pit. From there it is transferred to a furnace feed hopper, generally by an overhead crane. From the hopper the waste is fed onto moving mechanical grates. It burns as it moves continuously through the furnace. Noncombustible material falls from the end of the grate, is quenched, and then conveyed to trucks or temporary storage facilities. Ferrous metal may be recovered from this residue.

For the generation of steam, waterwall furnaces are enclosed by closely spaced water filled tubes. Water circulating through the tubes recovers heat radiated from the burning waste. Attached heat recovery boilers also generate steam while reducing the temperature and volume of the exhaust gases. Thus, these systems contain a variety of tube packages normally referred to as heaters, economizers, reheaters, etc. depending on the function of the particular zone. The marketable product, steam, is thus produced.

In the combustion process air is introduced into the furnace beneath the grates (underfire air) to aid in combustion and help keep the grates cool. More air is introduced above the fuel bed (overfire air) to promote mixing of the gases and complete combustion in the furnace. The combustion gases, after being cooled by passing through the various boiler sections, are cleaned by such pollution control devices as electrostatic precipitators and then vented to the atmosphere.

2. Processed Waste Utilization

Waterwall furnaces also are being designed to burn coarsely shredded MSW. By first shredding the solid waste, and possibly removing the ferrous metal and other noncombustibles, a more homogeneous and more controllable feed material can be produced. The shredded waste is fed into the furnace by spreader stokers which propel the waste across the combustion chamber to a traveling grate. Frequently this type of firing is referred to as semi-suspension firing. The waste is ignited while it is falling through the chamber, but combustion is completed while it rests on the grate.

One waterwall furnace combusting shredded MSW is in use in Hamilton, Ontario. Several others are in use for burning industrial wastes. Similar plants have been announced for Akron, Ohio, and proposed for Niagara Falls, New York. The Black Clawson Company has begun construction of a 2000 tpd unit in Hempstead, New York. That company will own and operate a semi-suspension furnace to burn a wet-pulped fuel produced from MSW by their patented process. These communities and industries have made the trade-off inherent in the use of process waste. The processed refuse is a more useful feed material for steam generating systems. The processing also permits recovery of materials which would otherwise be consumed or degraded in the burning process by recovering them prior to combustion. These benefits must be weighed against the higher costs of such systems.

B. Pyrolysis

The pyrolysis techniques being developed closely follow proprietary approaches under study and demonstration by private companies. As a result, the processes are known, most commonly, by the names of their developers. These technologies, however, can be categorized by the marketable fuel produced by pyrolysis. These fuels include low Btu gas, medium Btu gas and heavy oil.

1. Low Btu Gas: The Andco-Torrax System

The principal components of the Andco-Torrax System are the gasifier, secondary combustion chamber, primary preheating regenerative towers, energy recovery/conversion system, and the gas cleaning system. The solid waste is charged to the verticle slagging gasifier in as-received condition. The reactor vessel is designed so that the descending refuse burden and the ascending high temperature gases become a counter-current heat exchanger. The uppermost portion of the descending solid waste serves as a plug to minimize the infiltration of ambient air.

As the solid waste descends, three distinct process changes occur. The first is drying where moisture in the MSW is driven off. The second is pyrolyzing due to the heat transfer, and it is here that gas production occurs. The third is combustion in the vessel hearth where carbonaceous char is oxidized to carbon dioxide, and the noncombustible fraction is melted into a slag.

The heat for pyrolyzing and drying actions, and for melting the noncombustible fraction, is produced by the combustion of the char. This is accomplished with 2000°F preheated air supplied to the hearth of the gasifier. While the CO_2 rises into the gasifier to promote pyrolysis, the molten slag is drained continuously through a sealed tap into a water quench tank to produce a sterile, granulated residue.

The Btu content of the gas is in the 100-150 Btu/scf range, too low to make off-site transportation economical. Thus the gases are transported to an after-burner or secondary combustion chamber and burned to completion. The heat thus released is directed to a waste heat boiler and recovered as steam.

A portion of the hot waste gas from the secondary combustion chamber (about 15%) is directed through regenerative towers and the sensible heat is recovered and used to preheat the process air supplied to the gasifier hearth. Hot products of combustion from the after-burner and ambient process air pass through the towers for preheating the process air to 2000°F combustion air. The remainder of the secondary combustion chamber flow is supplied to the waste heat boiler designed for inlet gas temperatures of 2100 to 3000°F. The cooled waste gases from the regenerative towers are combined with the exiting flow from the waste heat boiler and are ducted to a hot gas electrostatic precipitator of conventional design.

2. Medium Btu Gas: The Purox Process

The key difference in the production of low and medium Btu gas from coal is whether the reactor vessel is fed air or oxygen. So it is with MSW pyrolysis. The Purox Process developed by Union Carbide Corp. utilizes oxygen rather than preheated air at the hearth of its counter-current vertical shaft reactor [9]. In opting for the production of 350 Btu/scf gas, Union Carbide favored the installation of oxygen production systems over the after-burner and regenerative tower systems of the Andco-Torrax system.

3. Oil Production: The Occidental Process

The Occidental "flash pyrolysis" process uses two stages of shredding, air classification, magnetic separation, drying, and screening to produce a finely divided fluff. This fluff is the feedstock for the pyrolysis reactor; it comprises about 60% of the incoming MSW. The fluff is fed, along with hot char, into a vertical stainless steel reactor. The hot char, the solid carbonaceous residue remaining after pyrolysis, provides the energy required to pyrolyze the incoming fluff. The material exiting the reactor consists of a mixture of char, ash, and pyrolysis gases. By rapidly cooling the gases before they completely react, a portion of the gas is condensed into a heavy oil fuel. The remaining gas, like the char, is used within the system.

By utilizing the elaborate feedstock preparation system, a byproduct is left which is high in glass and aluminum. The glass is extracted and purified by froth flotation. Eddy current devices recover the aluminum.

C. Mechanical Fuel Preparation Processes

The technology for preparing a supplemental refuse derived fuel (RDF) from MSW originated with the demonstration project supported in St. Louis, Missouri, by the U.S. EPA, the City of St. Louis, and the Union Electric Co. In early 1973 operations began where MSW was coarsely shredded prior to magnetic separation of the ferrous metals. The nonferrous material, including all combustibles, then was conveyed to steam generating boilers operated by the Union Electric Co. Shredded solid waste was test burned with coal. Significant deterioration resulted in the pneumatic feed system for blowing the shredded MSW into the boilers. As a result, the process was altered to include an air classification step after first stage shredding. An extended demonstration was undertaken with the shredded, air-classified MSW being burned in quantities providing as much as 20 to 25% of the heat input for the boilers.

Encouraged by this project, the Union Electric Co. formed a wholly owned subsidiary, The Union Colliery Co., from which plans were announced to build an 8000 tpd solid waste utilization system. Additional projects began to be formulated throughout the country, as Table II demonstrates. While some of these systems utilize the minimal processing employed in the St. Louis demonstration project, others employ more sophisticated process trains including secondary shredding of the air classified material. Still others chemically treat or briquette the secondary shredder product to produce a highly refined solid fuel. The popularity of this approach can be seen from analysis of Table II. It permits varying the capital input as the requirements of the fuel customer dictate.

Attention is being focused now on the utilization of trommels--rotating screens--to process RDF. The purpose of the trommeling step is to remove much of the rock, dirt, glass, and other heavy abrasive material prior to size reduction of MSW. Trommeling provides for reduced wear on the equipment, along with increasing the marketability of the recovered glass. Other component systems for extracting aluminum and other materials after air classification are not fully developed. The Milwaukee,

Wisconsin, and Chicago, Illinois projects listed in Table II exemplify the approach to these systems. Each project shreds the solid waste in two stages. Both projects recover ferrous metals. Both projects may install systems to recover glass, aluminum and other nonferrous metals at some later time should such unit processes become technically and economically feasible.

D. Gas Recovery From Landfills

Methane gas is produced continuously in waste disposal sites as solid wastes undergo decomposition. Consequently landfills can be viewed as potential fuel sources. The recovery of this methane-rich gas is feasible. The economic viability of existing demonstration and test programs, however, has yet to be proven. The recovery of gas is accomplished through wells. Pacey reports six methods of utilizing this landfill gas [7]:

(1) injecting this medium Btu gas (e.g., 500 Btu/scf) into an existing natural gas transmission line;
(2) delivering the medium Btu gas to an adjacent interruptible gas customer;
(3) using the raw landfill gas for on-site generation of electric power;
(4) treating the raw landfill gas, on site, to produce pipeline quality substitute natural gas;
(5) converting the raw landfill gas, on site, to methanol; and
(6) converting the methane to liquefied natural gas.

The first two process options are lower in capital cost. The landfill gas, 50% methane and 50% carbon dioxide, may be sold and used as is. The third method, on-site power generation, is currently being demonstrated by the Los Angeles Department of Water Power and the Department of Public Works at the Los Angeles Sheldon-Arleta landfill. This project uses the raw landfill gas as fuel for a 300 horsepower internal combustion engine driving a 200 kilowatt generator. The energy produced is distributed through an existing subtransmission system, supplying electric power for about 350 residences.

The fourth option requires construction of a processing facility to develop almost pure pipeline-quality methane. It necessitates removing the carbon dioxide, water, hydrogen sulfide, and other contaminants from the landfill gas. Processes in use or possibly suitable for gas purification include molecular sieves and an amine system. Among the systems for gas cleanup available are the following: Selexol, Fluor Solvent, Purisol, Rectisol, Benfield, Sulfinol, and Pressure Swing Adsorption (PSA). Collins [10] reports that PSA appears to be most cost effective at volumes of less than 10 million cubic feet of gas per day. Further, he reports that it has the advantages of being a dry adsorbant process and less mechanically intensive than other available processing methods.

The final two utilization systems--conversion of the gas into methanol or LGN--are costly, and likely suited only for landfills with very high gas production rates.

IV. ASSESSING THE ENERGY POTENTIAL FROM MSW

How much can, or will, MSW contribute to the energy supply of the U.S. in 1985 and the year 2000? Technically, energy can be recovered from solid wastes. However, the implementation of the many emerging technologies previously described necessarily depends upon other non-technical considerations such as product marketability, economic viability, public policy and many local demographic and political influences. It is important, particularly for planning purposes and establishing public policy options, to consider carefully the theoretical and realizable potential of energy from MSW.

Table IV presents three estimates: a theoretical upper bound, a high practical estimate and a low practical estimate. The waste tonnages presented in this table were taken from Table I. Waste growth rates from 2% (low) to 10% (theoretical max) were used to determine the annual waste tonnage levels for 1985 and 2000. The assumed percentages for wastes thought to be concentrated in urban areas are dependent primarily upon the population concentration and hence results in the "wastes available" for central processing facilities or large landfills. From there, general factors were applied to determine the amount available for energy recovery, and the percent of MSW which can be converted into useful energy. Similar steps were taken with methane generation to approximate the attention paid to enhancing or encouraging recovery from landfills [6].

The data presented in Table IV were derived from the assumed annual tonnages generated throughout the United States and projected for the years 1985 and 2000. The critical variable is the percentage of waste assumed to be concentrated in urban areas. It is from these wastes that the potential energy projection was derived. It is significant to note that the high practical estimates and low practical estimates for the year 1985 and the year 2000, as those estimates relate to energy recovery systems or landfill methane recovery systems, are not additive. For example, the low practical estimate in the year 2000 shows 0.8 quads for centralized resource recovery plants and 0.2 quads for landfill methane recovery systems. Those two estimates cannot be added to equal a quad. Rather, it must be observed that the higher of the two numbers in these tables (in this case the 0.8 quads) establishes the upper bound.

The mix of systems between centralized resource recovery plants and landfill methane recovery plants determines the extent to which that potential will be achieved as well as how that potential will be achieved. As can be seen, however, municipal

TABLE IV. National Estimates of Energy Potential From Solid Wastes for 1985 and 2000.[a]

	Annual wastes tonnage 1975 (millions/yr)	Waste growth rate (%/yr)	Animal wastes tonnage for subject year (millions/yr)	% Concentrated in urban areas
1985 Potential				
Theoretical maximum	625	10	1625	95
High practical estimate	427	5	696	75
Low practical estimate	164	2	200	50
2000 Potential				
Theoretical maximum	625	10	6791	100
High practical estimate	427	5	1447	80
Low practical estimate	164	2	269	60

TABLE IV. (Continued)

	Energy recovery systems potential			
	% Available for energy recovery	% Processable	Btu per pound of refuse	Energy equivalent (quads)
1985 Potential				
Theoretical maximum	100	100	5000	15.4
High practical estimate	60	75	5500	2.6
Low practical estimate	30	50	6500	0.2
2000 Potential				
Theoretical maximum	100	100	5500	74.7
High practical estimate	80	80	6000	8.9
Low practical estimate	60	60	7000	0.8

TABLE IV. (Continued)

	Landfill methane recovery potential			
	% Wastes landfilled	*Methane generation rate (cu.ft/lb)*	*% Methane Recoverable*	*Energy equivalent (quads)*
1985 Potential				
Theoretical maximum	*100*	*6*	*100*	*18.5*
High practical estimate	*95*	*4*	*75*	*3.0*
Low practical estimate	*75*	*2*	*50*	*0.2*
2000 Potential				
Theoretical maximum	*90*	*6*	*100*	*73.3*
High practical estimate	*75*	*4*	*80*	*5.6*
Low practical estimate	*50*	*2*	*60*	*0.2*

a. Caution should be exercised when deriving national estimates. The national estimates necessarily depend upon assumptions and judgments made by the estimator. This table reflects judgments made by the author for essentially three types of projections: theoretical maximum, high practical, and low practical estimates. The theoretical maximum estimates reflect an "upper bound" and are not indicative of what is feasible or realizable. The high and low practical estimates represent what may be a range of practicality for extracting energy from solid wastes.

wastes can contribute, on the *low* side, at least 0.8 quads in the year 2000. Assuming a high percentage growth rate in waste generation and the high concentration in urban areas, municipal wastes might possibly contribute up to 9 quads of energy in the year 2000 as a high estimate. Finally, it should be cautioned that these are projections of what may be done rather than forecasts of what probably will be done.

In conclusion, then, the practice of disposing of municipal waste by centralized resource recovery or by landfilling with concomitant methane recovery can contribute a significant amount of energy to the U.S. economy in the years to come. Some technologies are now commercialized and can be deployed. Other technologies are near the state of commercialization and will soon complement those already in existence.

REFERENCES

1. R. J. Black, A. J. Muhich, A. J. Klee, H. L. Hickman, Jr., and R. D. Vaughan, The National Solid Wastes Survey, An Interim Report, U.S. Dept. of Health, Education, and Welfare, 1968.
2. The Private Sector in Solid Waste Management--A Profile of Its Resources and Contribution to Collection and Disposal, Vol. 2, Analysis of Data, Applied Management Sciences, Inc. U.S. Environmental Protection Agency, SW-51d.1, Washington, D.C. 1973.
3. Frank A. Smith, Comparative Estimates of Post-Consumer Solid Waste, U.S. Environmental Protection Agency, Washington, D.C. SW-148, May 1975.
4. Third Report to Congress Resource Recovery and Waste Reduction, U.S. Environmental Protection Agency, Washington, D.C. 1975.
5. L. L. Anderson, Energy Potential From Organic Wastes, A Review of the Quantities and Sources, U.S. Bureau of Mines, Information Circular 8549, 1972.
6. J. R. Greco, "Energy Recovery From Solid Wastes--Considerations for Determining National Potential," National Solid Wastes Management Association, Technical Bulletin, Vol. 6, No. 10, November 1976.
7. John Pacey, Methane Gas in Landfills: Liability or Asset?, Proceedings of the Fourth National Congress on Waste Management Technology and Resource and Energy Recovery, U.S. Environmental Protection Agency, SW-8p, 1976.
8. J. R. Greco, "Resource Recovery From Solid Wastes--A Progress Report," National Solid Wastes Management Association, Technical Bulletin, Vol. 7, No. 5, June 1976.

9. David A. Tillman, "Mixing Urban Waste and Wood Waste for Gasification in a Purox Reactor," Thermal Uses and Properties of Carbohydrates and Lignins, Academic Press, 1976.

10. Robert H. Collins, III, Gas Recovery: National Potential, Proceedings of the Fourth National Congress on Waste Management Technology and Resource and Energy Recovery, U.S. Enviromental Protection Agency, SW-ip, 1976.

ENERGY FROM WASTE MATERIALS

1977 OVERVIEW

M. D. Schlesinger

Pittsburgh, Pennsylvania

I. INTRODUCTION

Interest continues at a high level concerning the disposal of organic solid wastes and where possible, recovering energy or materials that can be recycled from those wastes. New installations are coming on stream and are being monitored carefully. As a result, information is being published in the journals of professional societies, publications dedicated to solid waste management, books, and reports by government agencies. Some of these sources are reviewed here in an overview of problems and approaches.

The event that could have a very significant impact on all phases of solid waste management is the passage of PL 94-580. On October 21, 1976 Congress passed the Resource Conservation and Recovery Act of 1976, a bill that provides funds and establishes a major position within the U.S. Environmental Protection Agency to administer the provisions of the act. Several highlights of the act and their potential impact are discussed in this paper.

A. The Volume and Composition of Waste in the U.S.

The physical volume of waste is enormous. Its impact on the ecology is of great concern. Several studies estimate that the amount of solid waste generated per person per day in this country varies from 3 to 5 pounds [1,2]. Although the total appears large, estimates of the usefulness of this waste must take into account the source of waste, its distance from the markets for energy or recyclable materials, and its heating value. Volumes and compositions of waste are site-specific. Wide variations occur from one season of the year to the next. The level of industrial activity in a specific area also influences waste composition and volume. Although much literature exists concerning the production of feed, gas, or products of fermentation [3] from waste, disposal practices are only beginning to change. Most waste goes to landfill, while some is incinerated. Resource recovery is increasing as landfill space becomes less available and the economics of waste disposal change [4].

As the consumption of energy neared peak levels in the 1970s and a reasonable estimate could be made of remaining fossil fuel reserves, the need for increased recycling became obvious. Since most municipal waste in the United States is cellulosic in nature, it represents a source of energy whether it is combusted or converted into liquid or gaseous fuels. It is likewise realized that recovery of the metals and the glass for recycling represents a significant savings of energy. Aluminum is one example. Primary aluminum ingots require 244 million Btu/ton. Recycled aluminum requires only about 32 million Btu per ton, achieving a saving of 62,000 kwh/hr per ton of aluminum [4,5].

Estimates of the resource values in municipal waste vary considerably. An idea of the range, abstracted from several publications can be seen in Table I. Composition varies with location and many other factors. Moisture content of the as-received waste averages between 25 and 35%. The as-received material has a heating value between 4000 and 5000 Btu/lb. Methods of analysis are described in a recent publication [22]. In general, the methods used for the analysis of coal are applied to the wastes.

B. The European Comparison

The composition of the waste in several European cities was recently reported by Dr. Alter [6]. Many of the results are in the range reported in Table I. A few notable exceptions are

TABLE 1. Range and Composition of Municipal Solid Wastes

	Weight percent
Combustibles	
Paper	*35-60*
Garden wastes	*2-35*
Food	*2-8*
Cloth	*1-3*
Plastics	*1-2*
Noncombustibles	
Metal	*6-9*
Glass	*5-13*
Dirt	*1-5*
Moisture	*20-40*

evident. The percent of discard paper in most cities is below that of the U.S. Only Stockholm and Vienna fall in the same range as the U.S. Similar data on the range of composition of waste is presented by Fritz and Szekley [7] for European and North American municipal refuse. The European community countries produce about 1.7 billion metric tons of waste annually and this amount is growing some 5% per year [8]. In the Alter article, a description is given of the present European commercial, experimental, and research installations.

II. APPROACHES TO THE PROBLEM

An overview of energy from wastes was presented by Mr. Tillman at the New York meeting of the American Chemical Society, Division of Fuel Chemistry in April 1976 [9]. An expanded paper was prepared for the Federal Energy Administration [10]. Besides the components that make up municipal waste, the conversion systems for energy recovery were described. Another relevant study was made by the Bechtel Corporation for the Electric Power Research Institute [1]. Following is their tabulation of suppliers offering systems for converting solid waste to fuel:

Preparation as a solid fuel	23
Pyrolysis to a fuel gas	10
Pyrolysis to a fuel oil	1
Anaerobic digestion to a fuel gas	2
Combustion/gas turbogenerator	1
Incineration with heat recovery	30

Others will enter the field as the technology develops and installations prove themselves in practice.

A. System Selection

In the U.S., several plants for the conversion of municipal waste to energy exist. Three generate electricity and eight produce steam. Together they recover the energy from over two million tons of waste a year. Additional capacity is under construction and new plants are being announced regularly. Of the processes available, incineration to produce steam is the simplest method. There are several mechanical means for controlling the flow of waste within the incinerator; these account for the number of systems available.

Of the several other processing methods available, gasification by pyrolysis at high temperatures possibly provides the greatest flexibility. Gasification can be accomplished using air or oxygen for supplying the reaction heat by partial combustion of the waste. Low Btu gas ($\sim$ 100 Btu/ft^3) is produced when air is used. This gas is useful for industrial heating. It must be used hot to conserve the sensible heat in the gas. The gas produced when oxygen is used has a heating value of about 350 Btu/ft^3. Hydrogen and carbon monoxide from this fuel gas (or synthesis gas) can be used in a number of catalyzed reactions to yield gaseous hydrocarbons, liquid hydrocarbons, alcohols, or other chemicals.

From a practical standpoint, the potential use of waste disposal systems that recover energy is limited to three choices: (1) a water wall incinerator; (2) cocombustion of waste with a fossil fuel; or (3) pyrolysis to a combustible oil or gas. The first two have been demonstrated on a commercial scale. The other one has been demonstrated in process development units consuming about 200 tons/day, about half the size of a commercial module in some cases.

One assessment of advanced technologies was issued in January, 1976. The study was made at Columbia University under a grant from NSF/RANN, hoping to assist public officials make critical evaluations in selecting cost-effective and environmentally acceptable systems [16]. Nine processes were evaluated, these being representative of advanced technology and established operational reliability on an adequate pilot plant scale. They met environmental quality standards and were supported by a responsible industrial vendor. An estimate was made on each plant using a standard size (1000 TPD) and fixed operating, utility, and expense costs. Recovery credits were also standardized. Table II summarizes the costs and credits.

Table II. Basis of Cost Estimates

Costs	
Manpower (including all benefits and overhead)	*$20,000/man-yr*
Power	*3¢/kwh*
Other energy (oil, gas, steam)	*$2.10/MM Btu*
Maintenance	*2.7-6% of investment*
Materials and supplies	*Variable*
Landfill	*$6.00/ton*
Utilities	*Variable*
Insurance, taxes, legal	*1% of investment*
Credits, $/ton	
Ferrous scrap	*25*
Aluminum	*250*
Cullet (glass)	*15*
Frit (glass)	*3*
Sludge (dry)	*50*
Steam	*$2.30/MM Btu*
Electricity	*2¢/kwh*

All processes considered have certain features in common. Some separation can take place at an early stage such as the removal of clean cloth and bundled paper. These two items periodically have a good market value; if not, they are sent to the shredder and air classifier. The underflow from the classifier goes through magnetic separation, dense media separation, linear motor separator, an electrostatic separator and/or color sorters for the glass. Concentrated organic waste then goes to combustors or reactors. After products are recovered, the residue is made acceptable for landfill. These processes are capital intensive, closely coupled unit operations. The obvious risk is great when the plant scale is large. Either a sizeable buffer capacity is required or the entire plant is shut down when a single unit has a problem. First or second generation systems, particularly, are often not completely reliable. Mechanical scaleup can be uncertain, especially when the system operates at severe conditions and/or unproven components must be installed (e.g., large valves).

Practically all unit operations are applied to the processes that have been described. Where they are placed in the process train is a matter of selection by the engineer and designer. Selected descriptions are given in other publications [1,16,13] and in the paper by Mr. Greco. They may be generalized as shown in Fig. 1. Typical variations include consumption of the processed fuel in another process.

A summary of the comparative study by Benziger, et al [16] is shown in Table III. The cocombustion process seems to be the most cost-effective. It is also highly energy efficient. The most costly process appears to be the anaerobic digestion of municipal waste because of its high operating cost. Its energy efficiency is also lowest of the nine systems studied. The four systems now shown in the table are intermediate in operating cost and efficiency.

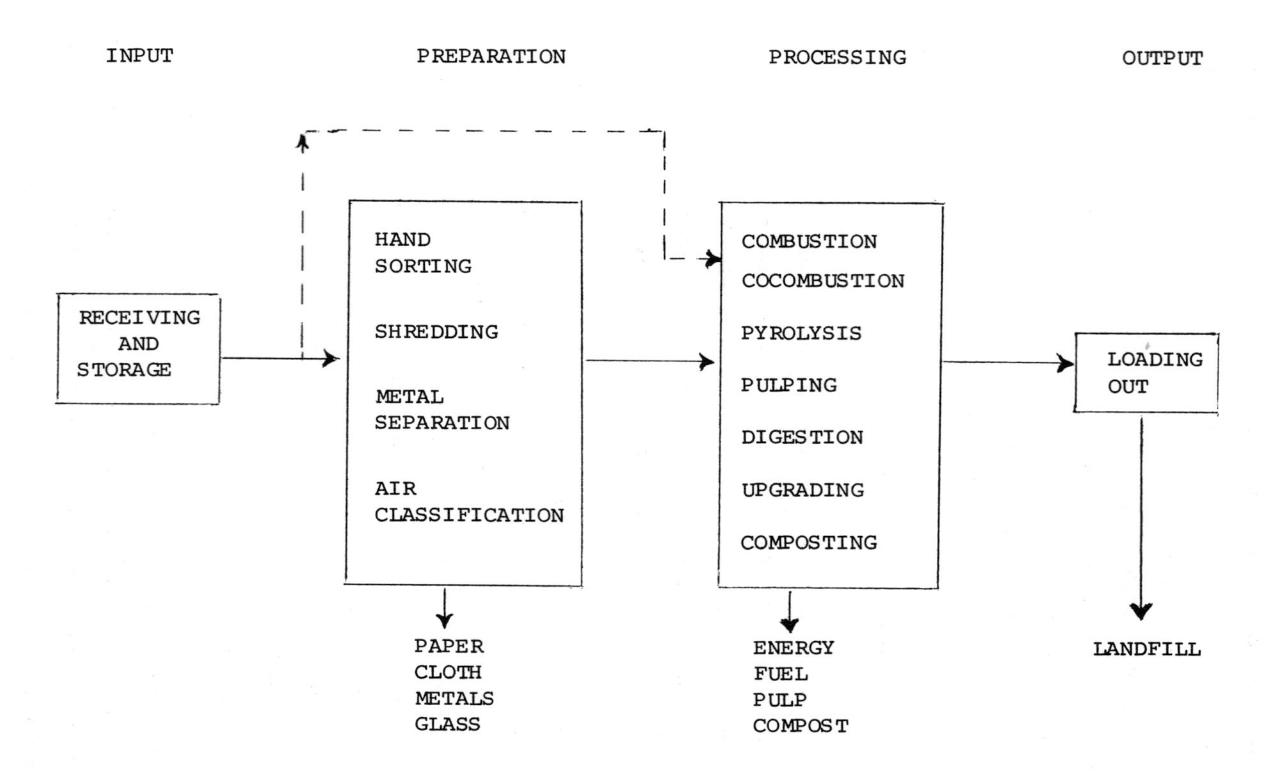

FIGURE 1. Generalized Processing Scheme for Solid Waste.

TABLE III. Energy Recovery Processes (Basis 1000 TPD) [16]
(Costs in $/ton)

	Incin-eration	*Co-combustion*	*Pyrolysis Gas*	*Pyrolysis Liquid*	*Digestion*
Investment	*$30,800*	*$10,400*	*$21,500*	*$21,900*	*$19,600*
Operating Cost	*11.13*	*6.36*	*12.00*	*12.31*	*18.42*
Credits	*15.48*	*8.29*	*13.10*	*9.28*	*11.42*
Net disposal cost	*5.89*	*1.53*	*6.07*	*10.31*	*13.53*
Energy efficiency	*67%*	*66%*	*44%*	*37%*	*25%*

It should be pointed out that the credits shown are not always possible. Either the potential users are not with an economic transportation area or the amount of recycle material exceeds the amount that the market can handle thus depressing the potential income.

B. The Decision Making Process

Handling municipal wastes is as much a social problem as an economic one. After the cold winter of 1976-77, more people will expect and, in some cases, demand that all possible energy be saved. Local governments could be hard pressed to make decisions about the selection of the proper waste disposal and/or energy recovery system for their area. Problems could occur because of the advanced technology involved plus the related problems of finance and system management. The capital costs of recycling are high. The choice of a disposal system might have to be made from unproven alternatives.

Warnings continue concerning the need to do something about the wastes discharged to the environment. The admonitions concern the deepest oceans and caverns of the earth to the stratosphere. Concern is for the impacts on resources, places to dispose of waste materials and the potential destruction of the environment. Perhaps we can no longer wait. The time may be here to make a selection from today's best systems.

C. Assistance for Making the Decisions

One problem voiced by designers is that some public officials are vague when indicating what they want a waste disposal-energy recovery system to accomplish. The engineering company cannot prepare a responsive bid and hence problems of evaluation can occur. Conversely, in order for the system presented by a bidder to appear attractive, the bidders may estimate unrealistically low values for the engineering and operational costs coupled with high values for the recovered materials. These discrepancies can be large and most obvious to the experienced estimator or one who has worked through a detailed request for bids. For example, in one plant [11], paper recovery was assigned a value of $214 million per year. The actual value was only $29 million. A similar discrepancy was found in the non-paper residue: $200 million versus $11 million. Furthermore, the land requirements were excessive and the construction time was too long. The net effect of reevaluation was a multimillion dollar saving.

The chemist or chemical engineer can provide significant support as a member of a task force or as a consultant. His specialized knowledge can help as the design evolves. These inputs can provide city engineers and officials with insight into the system under consideration. The result can be the formulation of more precise questions for the vendors, and thus the purchase of a better recycling system. Insight into the technologies also helps officials feel comfortable that they can make the best decision regarding system selection. A challenge to municipalities is to make an intelligent choice from numerous alternatives, finding the best one for their situation [12,13,14].

One barrier to energy recovery was reported in a local newspaper on March 2, 1977. The article recalled that the mayor had stated that he was convinced that incineration with a waste heat recovery system was the best plan for the city. That was two weeks before he planned to run for the U.S. Senate. Now, two and a half years later, the city administration announced that they do not think $45 million should be spent for the plant. The city purchased new compactor trucks that can achieve double the loading capacity of older trucks. The new trucks with their eight ton load can go directly to the landfill site at a savings of about 70% when compared to the same eight tons delivered to a transfer station.

This lower cost has reduced the urgency of complying with the plan that the city filed with the state environmental department. The long range plan calls for the conversion of solid waste to energy. Two other factors, however, come into the local planning: the continuing availability of landfill sites and the social value of creating new energy supplies. In the situation cited, there is landfill capacity for another 50 years.

III. OTHER MUNICIPAL WASTE MATERIALS

Sewage sludge is a solid waste that is difficult to dispose of safely. Furthermore, in many plants it is not accumulated in large quantities. About one ton of dry sludge is recovered for every million gallons of waste water. Of the more than 22,000 municipal treatment plants, less than 350 of them are designed for more than 10 million gallons per day and about 15,000 are designed for less than one million gallons per day. Most of the smaller plants and even some of the larger ones, sell or give away the digested or otherwise treated sludge. Sewage sludge is an excellent soil conditioner but it has a low nutrient value. Further, the sludge used in agriculture can contain pathogenic agents or heavy metals unless these contaminants are controlled. One approach to the reduction of pathogen levels in waste has been irradiation of sludge using gamma emitters or an electron accelerator [17].

Composting is another approach. Pathogenic agents, except for salmonella, are removed by composting fresh sludge mixed with composted sludge. Even salmonella is reduced in the warmer sections of the compost pile where the temperature is lethal to the salmonella. Odor problems are much less with some composting than they are with other forms of disposal. Passing gases from a compost pile through a pile of finished compost almost eliminates any odor [17]. Composting of partially dewatered sludge can decrease its water content from about 75% to 40%. If this composted sludge is then fired in an incinerator, the use of auxilliary fuel might be reduced or eliminated.

IV. AGRICULTURAL AND FORESTRY WASTES

Combustion of bagasse and some other agricultural wastes for the generation of energy is practiced widely. Although the unit heating value may not be as high as equivalent weights of natural gas or petroleum, there is a negligible raw material cost, only that necessary to move it from the mill to the boiler. The field collection of wastes and transportation to a central point for combustion is usually energy deficient, that is, the energy consumed is greater than the energy in the load. When agricultural wastes are used for steam raising, it is necessary for the combustion zone to be designed for the particular fuel being fired. Account must be taken for the heat release, the moisture content of the feed, its particle size, the need for supplemental fuel, and ash removal. Other factors may also be significant in particular cases.

Bagasse is an appropriate example. Normally it contains 40 to 55% moisture after it is pressed and ready for disposal. Its moisture-free chemical composition averages 43% carbon, 6% hydrogen, 47% oxygen and 2% ash. The moisture ash-free higher heating value is about 8500 Btu/lb. Since steam generation is considered, the heating value of bagasse is only about one-half that of an equivalent weight of coal after correcting for the moisture and ash. The composition of bagasse is about the same as green wood removed from the forest, high in moisture, and with a chemical formula close to cellulose $(C_6H_{10}O_5)$.

Spreader stokers are preferred for bagasse and coarse wood waste. Travelling grate stokers are used often for coke breeze and some other large particle wastes. Fine materials such as sawdust, rice hulls, coffee grounds, and chars are usually tangentially fired alone or in combination with supplemental fuels. In a dual fuel boiler, the wet materials are injected high in the combustion chamber where they dry before combustion occurs. All but the largest particles burn before reaching the grate at the bottom of the furnace.

Feedlot wastes may contain 80% or more of water. The food scrap component of municipal waste and food processing wastes also has a very high moisture content. Unless the above types of materials are air dried, there is not enough heating value in the combustible portion to heat and evaporate the moisture content. Open air drying of such materials can be objectionable.

A recently developed process, COSTEAM, can handle waste materials with a high moisture content [18]. In January 1977 an experimental facility was dedicated at Albany, Oregon [19]. Wood chips will be the initial feed stock but such wastes as sugar beet tops, corn stalks or similar wastes can be charged. The three-ton-per-day plant was designed by Rust Engineering Co. and is being operated for ERDA by Bechtel, Inc. The process can react cellulosic materials with carbon monoxide or with synthesis gas ($CO + H_2$) to produce about two barrels of oil per ton of dry feed. Operation is at 380°C and 100 to 250 atmospheres pressure. The oil product has a heating value of 15,000 Btu/lb (a little lower than petroleum derived fuel oil) because of its high content of oxygen. The oil is low in sulfur, about one-half the present EPA limit for fuel oil.

V. FEDERAL ASSISTANCE

Congress passed Public Law 94-580 on October 21, 1976. The act is "to provide technical and financial assistance for the development of management plans and facilities for the recovery of energy and other resources from discarded materials and for the safe disposal of discarded materials, and to regulate the management of solid waste." Congress found that solid wastes were being generated at an ever increasing rate, and that the Federal government should provide assistance and leadership in solid waste disposal practices. The nature of the problem is expressed in the statement that many cities will be running out of suitable waste disposal sites in the next five years. The recovery of materials now discarded to landfills represents a means for protecting public health and the environment, and for conservation of domestically scarce resources. The conserved resources include equivalent amounts of fossil fuels and the energy expended in the refining of metals.

The objectives of the act are to promote the protection of health and the environment and to conserve valuable material and energy resources by:

(1) providing technical and financial assistance,

(2) providing training grants,

(3) prohibiting open dumping,

(4) regulating hazardous wastes,

(5) providing guidelines,

(6) providing research and development programs,

(7) promoting demonstration systems, and

(8) cooperating with Federal, State, and Local governments.

Subtitle B of the act requires that the Administration of EPA establish a post of Deputy Assistant Administrator within the Environmental Protection Agency. The Deputy Assistant Administrator will head the Office of Solid Waste, providing consultation and assistance to local agencies for the implementation of solid waste programs.

Of the general authorization to carry out the provisions of the act, at least 20% of the $35 million in 1978 will be spent for resource recovery and conservation panels established to assist localities. Similarly, at least 30% of the funds will be spent for the management of hazardous wastes.

The hazardous wastes portion of the act involves the identification and listing of materials considered to be hazardous plus the promulgation of regulations and standards regarding transportation, storage and disposal facilities; permit requirements; authorization of state programs; access to sites and records; and compliance. As is typical with many Federal regulations, no state or political subdivision may impose any requirements less stringent than those authorized in the act.

Also included in the act is an authorization for small grants to the states, assisting them in the development and implementation of authorized hazardous waste control programs.

Subtitle D is concerned with State or regional solid waste plans. Particular emphasis is expressed about sanitary landfills and open dumps. The stated objective is "to assist in developing and encouraging methods for the disposal of solid waste which are environmentally sound and which maximize the utilization of valuable resources and to encourage resource conservation." Federal assistance amounting to $30 million, authorized for Fiscal Year 1978, is available to help the States develop and implement their plans. In addition, there is another $42.5 million possible each year for technical assistance including aid to small or rural communities where most of the wastes dumped are from the outside, or when there are serious environmental problems.

The public law documents that land is too valuable a resource to be polluted by discarded materials; disposal of solid and hazardous wastes can present a danger to human health and the environment; the amounts of solid wastes has increased; and alternatives to existing methods of land disposal must be developed. Even when adequate landfill or dump sites are available, they will be phased out and new technology applied.

VI. CONCLUSION

There is no possibility that municipal or other solid waste could ever become a primary contributor to the total U.S. energy supply of this country. However, there are and will continue to be instances where all or some locally needed energy will be supplied by solid waste. For example, the wood waste at some mills can supply the energy to run the entire plant during the logging season. A system for year-round energy supply generally needs both solid waste and supplemental fuel.

Interest in energy from waste can be expected to continue at a high level. New plants now under construction will be coming on stream and new plants will be started as communities find the system that best suits their needs. Both public and private research laboratories will continue to investigate new methods of solid waste disposal. Considerable impetus can be gained through organizations such as the National Center for Resource Recovery, Inc., supported by labor and industry. They are dedicated to carry out research, provide training, and disseminate information related to the recovery of usefull components from municipal solid wastes. For several years public agencies have pursued an aggressive program of research on resource recovery. Leaders in the research on solid waste have been the U.S. Bureau of Mines, the U.S. Energy Research and Development Administration, the National Science Foundation, and the U.S. Environmental Protection Agency. The new Resource Conservation and Recovery Act should provide even more impetus to solving many of the remaining problems. A good start has been made. The future should see even more progress with new achievements in both the public and the private sectors.

ACKNOWLEDGMENTS

I wish to express my appreciation to Dr. Bernard D. Blaustein of the Pittsburgh Energy Research Center (ERDA), Professor M. A. Shapiro of the University of Pittsburgh Graduate School of Public Health, and David A. Tillman of Materials Associates for their careful review of the draft paper and their helpful suggestions.

REFERENCES

1. Bechtel Corp., "Fuels From Municipal Refuse for Utilities: Technology Assessment." EPRI 261-1 Final Report Prepared for The Electric Power Research Institute, Palo Alto, Calif., March 1975.
2. Terri Schultz, "Garbage Is No Longer Treated Lightly." New York Times, March 6, 1977.
3. Ken Pober and Hans Bauer. "From Garbage-Oil." CHEMTECH, March 1977.
4. Max J. Spendlove, "Recycling Trends in the United States." U.S. Bureau of Mines Information Circular 8711, 1976.
5. Battelle-Columbus, "Energy Use Patterns in Metallurgical and Nonmetallic Mineral Processing," Interim Report. Open File Report, U.S. Department of the Interior Library, Washington, D.C.
6. Harvey Alter, "European Materials Recovery Systems." Environmental Science and Technology, Vol. 11, No. 5, May 1977.
7. Jack J. Fritz and Julian Szelkey, "Refuse: Resource or Liability." CHEMTECH, April 1977.
8. "Coordinated Waste Management in Europe," Chemical and Engineering News International, March 14, 1977.
9. David A. Tillman, "Energy From Wastes: An Overview of Present Technologies and Programs," Fuels From Waste, Academic Press, 1977.
10. David A. Tillman, "The Contribution of Non-Fossil Organic Materials to U.S. Energy Supply." Federal Energy Administration, Washington, D.C., 1977.
11. "Dade County's Resource Recovery Plant," Resource Recovery and Energy Review, Wakeman-Walworth, Inc. Darien, Conn., Mar/Apr 1975.
12. Robert A. Colonna and Cynthia McLaren, "Decision Makers Guide in Solid Waste Management Programs," (SW 127) U.S. Environmental Protection Agency, Cincinnati, Ohio, 1974.

13. Steven J. Levy and H. G. Rigo, "Resource Recovery Plant Implementation Guide for Municipal Officials"(SW 157.2), Office of Solid Waste Management Programs, U.S. Environmental Protection Agency, 1975.
14. Alan Shilepsky and Robert A. Lowe. "Resource Recovery Plant Implementation" (SW 157.1). Office of Solid Waste Management Programs, U.S. Environmental Protection Agency, 1976.
15. Edward J. Farkas, "Research Directions in Solid Waste Management." Ind. Eng. Chem. Fundamentals,Vol. 16, No. 1, 1977.
16. J. B. Benziger, B. J. Bortz, M. Neamatella, R. M. Szostak, G. Tong, R. P. Westerhoff, and H. W. Schulz. Resource Recovery Technology for Urban Decision Makers. Urban Technology Center, School of Engineering and Applied Science, Columbia University, New York, 1976.
17. W. C. Remini, E. J. Wahlquist, and H. D. Sivinski, "Beneficial Use of Waste Nuclear Isotopes." ERDA 77-17, U.S. Energy Research and Development Administration, Washington, D.C., January 5, 1977.
18. H. R. Appell, Y. C. Fu, E. G. Illig, F. W. Steffgen, and R. D. Miller, "Conversion of Cellulosic Wastes to Oil." U.S. Bureau of Mines Report of Investigations 8013, 1975.
19. "Biomass/Synfuel Plant Dedicated," ERDA News, January 24, 1977.
20. "Biomass Conversion: Economics Inhibit Progress," Chemical and Engineering News, February 28, 1977.
21. William J. Jewel, Energy, Agriculture and Waste Management. Ann Arbor Science Publishers, Inc., 1975.
22. H. Schultz, P. M. Sullivan, and F. E. Walker. "Characterization of Combustible Portions of Urban Refuse for Potential Use As a Fuel." U.S. Bureau of Mines Report of Investigations 8044, 1975.

DISCUSSION OF CRITICAL ISSUES

The Editors

I. INTRODUCTION

Four questions dominated the general discussion: (1) the drying of biomass fuels prior to combustion, (2) the use of biomass fuels in non-integrated papermaking plants, (3) the environmental impacts of pyrolysis, and (4) the practicality of energy plantations. Additionally, Dr. Rosely Maria Viegas Assumpcao from San Paulo, Brazil, provided a presentation on biomass energy utilization in that nation.

II. THE DISCUSSION QUESTIONS

A. Predrying Wood Fuels

The question posed was: What is the best way to predry biomass residues to improve furnace capacity and responsiveness?

The consensus was that predrying is not economic and may present environmental problems. George Voss of American Fyr-Feeder Engineers stated that predrying creates added airborne emissions. It presents fire hazards when applied to wood. Thus, their firm does not recommend going beyond air drying. Dr. John Zerbe of the U.S. Forest Service stated that, initially, they believed predrying to be useful. They believed it would improve fuel value, reduce capital investment and aid in stack gas cleaning. On further investigation, they have concluded that for large-scale operations, burning without predrying makes more sense.

Two specific problems were pointed out. Dr. Larry L. Anderson of the University of Utah observed that drying causes low-grade oxidation and some loss of fuel value. Dr. David Brink of the University of California, Berkeley, stated that predrying causes the production of some volatiles which would have to be dealt with for environmental protection purposes.

B. The Use of Biomass in Non-Integrated Paper Mills

The question posed was: The non-integrated paper plants relatively distant from pulp mills have difficulty increasing their energy self-sufficiency. How can they improve their position?

Dr. Kyosti V. Sarkanen of the University of Washington stated that vertical integration of forest industries, including saw mills plus pulp and paper mills, is more prevalent in Scandanavia than in the U.S. Thus greater opportunity in the drive for energy self-sufficiency exists there.

Dr. John Grantham of the U.S. Forest Service and Dr. Anderson voiced the only practical option available: look to other sources of biomass, treating biomass fuels only as a locally available and usable energy source. This implies that a biomass fuel produced by one industry need not be used, necessarily, by that same industry.

C. Environmental Problems of Pyrolysis

The question posed was: Pyrolysis can produce soluble tars resulting in high biological oxygen demand (BOD) loadings on waste water treatment systems and surrounding water systems. How can this problem be solved?

It was observed that production of soluble tars is a function of temperature. Dr. Brink pointed out that pyrolysis systems operating at 400-700°C (750-1300°F) or fixed bed counter-current systems which pass pyrolysis gas through incoming feed material for drying purposes do generate relatively large volumes of soluble tars. High temperature systems (e.g., 1000°C) completely gasify the incoming biomass.

Dr. Sarkanen stated that this problem solution favors fluidized bed gasification. George Voss agreed that fluidized bed systems solve the problem of tar formation but have other disadvantages including low turndown rates (e.g., turndown rates in the range of 1.0 to 1.5). Thus he concluded that an inclined grate gasifier with a thin bed of material offers a better design solution.

D. The Potential for Energy Farms

The final question was: What is the potential for energy farms, particularly in the United States?

Much of the energy from biomass debate centers around the formation of energy farms. One paper presented at the symposium, Thermal Uses and Properties of Carbohydrates and Lignins (also published by Academic Press), dealt specifically with the growing and harvesting of crops for energy. The paper by Dr. Grantham in this symposium also dealt with energy farming.

Dr. Anderson stated that, while we know how much carbon is produced by photosynthesis anually, energy farming is a moot point in this country. We can wait for results from countries outside the U.S. who are either utilizing energy farming or about to begin using such systems.

Dr. Grantham pointed out that the problem of underutilized non-industry owned private commercial forest land is of higher priority in this country. He said that the price of fuel wood could help solve this problem. Dr. Brink added that understocked Federal forests also need attention. Dr. Grantham pointed out that 10 million acres of Federal forest lands are understocked, but cautioned that much of this land is dedicated to other objectives. The consensus was that, for the U.S., energy farming is not currently a practical option.

III. THE USE OF BIOMASS FOR ENERGY IN BRAZIL

The presentation by Dr. Assumpcao covered three specific areas: (1) the present use of wood and agricultural crops for energy, (2) the growth of biomass to support such energy production, and (3) the future use of biomass fuels and energy in Brazil.

A. The Present Use of Biomass Energy in Brazil

Table I presents the present sources of energy in Brazil, and their contribution to its economy. Significantly, vegetation based fuels supply 27.4% of that nation's energy needs while fossil fuels supply only 51.9%.

The biomass fuels come principally from wood, with lesser amounts coming from agricultural materials (e.g., sugar cane). Among the industries using such fuels are the steel industry, the chemical industry, and the pulp and paper industry. Brazilian steel mills use charcoal as a reductant. The chemical industry uses ethanol, produced principally from sugar cane (93%) with modest amounts (7%) coming from wood residues. The pulp and paper industry uses wood residue for fuel.

TABLE I. Energy Supply in Brazil, 1973

Source	*Contribution energy supply (in %)*
Biomass	*27.4*
Hydroelectric power	*20.7*
Oil	*48.4*
Natural gas	*0.3*
Coal	*3.2*
Total	*100.0*

In addition to industrial use of vegetation based energy, motor vehicles are fueled partially by such products. Currently ethanol and gasoline are blended in a 15%:85% mixture and this fuel is then used. Residential fuel wood is also used.

B. The Growth of Biomass

Supporting this use of renewable resource fuels is a vast forest resource. Some 870 square miles of Brazil are forested, an area representing 41% of the total land mass of that country. This forested land share of the nation compare to 32% in the U.S. Further annual biomass production is 400 ft^3/acre compared to 38 ft^3/acre in the U.S.

Some 77.5% of this forest land area is in the tropical forests of northern Brazil; 10.9% is in the southern forests. In the southern forests, which support much of the industrial activity, 60% of the growth is in eucalyptus trees while 37% is in pine species. These tree species have rotation times of 7 and 10 years respectively. The eucalyptus rotation for the pulp and paper industry is 9 years. Such rotation times are in stark contrast to a 40-year rotation time for U.S. forests.

C. The Future of Renewable Resource Fuels in Brazil

By building upon the present base, Brazil projects significant growth in the use of renewable resource fuels. Growth is planned in the industrial sector, as exemplified by the pulp and paper and steel industries, and in the use of ethanol as a transportation fuel. That country also plans major developments in the growth of vegetation to support renewable resource energy supplies.

The projected growth of the pulp and paper industry is shown in Table II. By the year 2000, Brazil projects that it will command 10% of the world capacity in this wood residue fueled industry.

TABLE II. Projected Production of Pulp and Paper in Brazil

Year	*Projected production (in million metric tons)*	*Projected share of world production capacity (in %)*
1975[a]	*1.66*	*1.3*
1980	*4.94*	*3.3*
2000	*32.50*	*10.0*

a. Actual

In the steel industry, charcoal now provides 29% of the fuel for iron ore reduction. It is expected to increase its role as a reductant to over 30% by 1980, and continue to increase beyond that point. In the production of motor fuels, Brazil projects going from a 15%;85% blend of ethanol to gasoline, to a 20%:80% blend.

Supporting such increases in renewable resource fuel utilization are several activities. Brazil is creating 30 Forest Districts, each containing two 1000 ton/day pulp mills. These forest districts will occupy 10.5 million acres--0.5% of Brazil's land mass.

Charcoal for steel production now requires 0.65 acres of sustained yield eucalyptus forest per ton of steel output. Brazil is reducing the harvesting rotation time from 7 to 5 years, recognizing that wood density controls the quality of charcoal produced. To support the increased demand for ethanol, Brazil is planning on an additional 2.3 million acres of planted sugar cane. This crop will be used for fuel production.

Because Brazil must import some 80% of the fossil fuel it uses, it has made great progress in the use of renewable resources, and plans to accelerate its utilization of such fuels.

INDEX

H

I

L

M

N

O

P

R

S

T

U

V

W

A B 7 C 8 D 9 E 0 F 1 G 2 H 3 I 4 J 5